EXPLODING TECHNICAL COMMUNICATION: WORKPLACE LITERACY HIERARCHIES AND THEIR IMPLICATIONS FOR LITERACY SPONSORSHIP

Dirk Remley
Kent State University

Baywood's Technical Communications Series
Series Editor: Charles H. Sides

Routledge
Taylor & Francis Group
LONDON AND NEW YORK

First published 2014 by Baywood Publishing Company, Inc.

Published 2017 by Routledge
2 Park Square, Milton Park, Abingdon, Oxon OX14 4RN
711 Third Avenue, New York, NY 10017, USA

Routledge is an imprint of the Taylor & Francis Group, an informa business

Library of Congress Catalog Number: 2013049919
ISBN 13: 978-0-89503-889-0 (hbk)
ISBN 13: 978-0-89503-890-6 (pbk)

Library of Congress Cataloging-in-Publication Data

Remley, Dirk.
    Exploding technical communication : workplace literacy hierarchies and their implications for literacy sponsorship / Dirk Remley, Kent State University.
        pages cm -- (Baywood's technical communications series)
    Includes bibliographical references and index.
    ISBN 978-0-89503-889-0 (cloth : alk. paper) -- ISBN 978-0-89503-890-6 (pbk. : alk. paper) -- ISBN 978-0-89503-891-3 (e-pub) -- ISBN 978-0-89503-892-0 (e-pdf)
  1. Technical communication--Case studies. 2. Workplace literacy--Case studies. 3. Technical communication--United States--History. I. Title
    T11.R4245 2014
    601'.4--dc23
                                                                2013049919

Table of Contents

LIST OF FIGURES

LIST OF TABLES

Acknowledgments

I wish to acknowledge the encouragement, support, and assistance of a few people as I worked on this study, especially the researching of it as well as some of the writing. Productive scholarship comes from dialogues with others about information one may collect or observe. Such conversations can occur at various states of the process, and they facilitate reflection. It is important to recognize those associated with such discussions and reflection.

I thank Brian Huot for his mentorship, encouragement, and support as well as for his patience with me. His background with literacy policies and scholarship helped frame the study within the field. Thanks, also, to Pamela Takayoshi, who provided many insightful comments and suggestions with research methods and helped others involved understand the way I was treating the topic of multimodality relative to the sponsorship of literacy. She encouraged me to reflect on my research methods not only relative to methodology scholarship but also relative to the study itself. Special thanks to Raymond Craig, who facilitated collection of some preliminary data for the study and analysis relative to multi-modal rhetorical theory.

I also thank those members of the Arsenal and community who helped locate sources and participated in the study. Their assistance and participation provided narratives to bring information from the archived documents to life. There were times when I sat in the library at the Arsenal and coded data from the archived documents dutifully, and there were other times when I sat in awe of the information presented about activities there. Without the assistance of others, I would not have been able to locate such information.

Acknowledgments

CHAPTER 1

Introduction

LITERACY AND DESTRUCTION

Bombs and howitzer shells exploded all over Europe and Asia during World War II, destroying buildings and killing millions of people—soldiers and civilians. This was a normal part of war. However, on March 24, 1943, in the midst of World War II, an explosion on the home front, in the Depot area of a United States arsenal that stored bombs and howitzer shells, destroyed a storage igloo and killed 11 people, injuring another 3. Of the 11 killed, 9 were incinerated by the blast.

Over 2,000 clusters of fragmentation bombs (Figures 1.1 and 1.2) exploded at once. The explosion destroyed an entire 60-foot long storage igloo of reinforced concrete (like that in Figure 1.3) as workers were unloading and moving boxes of the small bombs. Structures as far away as 2,100 feet (the length of almost seven football fields) from the epicenter of the blast were damaged. A subsequent investigation concluded that the combination of a defect in the fuse that was used to detonate the bombs and the rough handling by workers unloading the bombs for storage caused the explosion to occur (Stratton, 1943, p. 13).

Workers experienced hands-on training, in which safe unloading practices were encouraged. Further, workers were encouraged to handle the particular shipment carefully; a memo alerted administrators of the Arsenal that the fuse was defective. Nevertheless, the bombs were handled unsafely.

How is it possible that even with information available about a given defect that could cause so much damage and encouragement from managers and administrators to be careful, workers could mishandle bombs with catastrophic results? What or who is to blame for such an event? Information about the defect was known to management, and managers told workers to be careful, so it could not be management's fault. Further, workers went about unloading the bombs as they normally had; while it may have been rough, never had such an accident occurred until that moment. How would they have understood that the way they handled the bombs was "rough?" So answering the question of how it is possible needs a different approach.

1

Figure 1.1. Illustration of fragmentation bomb cluster.
Source: From "M1A1 Clusters of U.S. 20-pound (M1) fragmentation bombs."
U.S. Army Medical Department Office of Medical History. October 5, 2012.
Photo from U.S. Army Medical Department Office of Medical History Website.

One needs to ask questions like how were workers trained to handle bombs? What literacy practices at the workplace may have contributed to the accident? Could managers have done anything more to encourage workers to be careful? Could workers have done anything to help management understand that the way they handled bombs may have been considered too rough? These questions, like those in the preceding paragraph, pertain to technical communication practices; however these questions pertain more to *how* people in the workplace communicated about certain tasks and across different levels of the organization, not just whether there was communication. That is, what literacy practices and skills were associated with the way people communicated, and why were these forms of literacy used by different people in the organization?

Several variables are involved in organizational communication practices, and this book discusses a number of attributes of the communication and literacy practices that occurred at the Arsenal and why they occurred as they did. The training workers experienced was developed by the federal government and industry leaders to minimize the impact that relatively low literacy rates at the time would have on training people in war industry work. It emphasized visual forms of communication to teach procedures. However, this accident helped to

BOMB, FRAGMENTATION, 20-LBS, AN-M41 & AN-M41A1

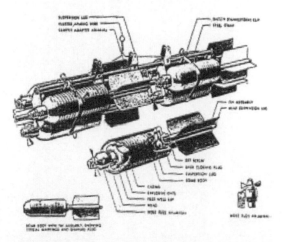

Body. This bomb is constructed of cast-steel nose and tailpieces, a seamless steel inner tube, and a helically wrapped drawn steel wire wrapping around the inner tube. The tube is threaded to hold the nose and tail section

Suspension. For individual suspension of this bomb, a U-shaped eyebolt of steel is welded to the body at the center of gravity for horizontal suspension, and an eyebolt is welded to the tail for vertical suspension. The bomb may be dropped in a cluster of six bombs in the *Cluster Adapter AN-M1A2 or M1*, forming the *Cluster AN-M1A1 or M1*. The cluster adapter is made of sheet steel, and does not use eyebolts of bombs for suspension.

Tail. Four rectangular sheet-steel vanes are welded to a length of one-inch cast-iron pipe which screws into the base-filling plug.

Over-all length	19.5 inches
Body length	11.3 inches
Diameter	3.6 inches
Over all weight	20.3 pounds
Filler	TNT
Filler weight	2.7 pounds
Fuzing	M158, AN-M110A1, M110, M109

Figure 1.2. Description and specifications for fragmentation bomb.
Photo courtesy of the U.S. Army.

identify a problem with the training and different literacy practices across different levels of the organization. Consequently, it would affect technical communication documents and practices, especially those associated with training, at installations around the country as people attempted to address literacy differences across various levels of the organizational hierarchy. Through an

Figure 1.3. Interior of storage igloo. Photo by Dirk Remley.

investigation, it became evident that the different literacy practices at the Arsenal may have contributed to causing the accident. I discuss the conflicting dynamics within the training and other technical and managerial communication materials that contributed to this catastrophic accident and subsequent changes to these technical documents and other communication practices to address those issues. Indeed, the changes that were made to the technical communication practices especially ensured that such an accident would not happen again.

However, the particular practices that contributed to the accident were influenced by a number of factors nationally and locally. Because literacy practices are culturally bound and context-dependent, in this book I describe an entire ecology that contributed to the literacy practices at the Arsenal as well as those practices outside of the Arsenal; that is, I describe how practices at the Arsenal affected practices outside of it. Indeed, while a variety of literacies were practiced at the Arsenal, it was print-linguistic literacy that was favored over other forms, and that favoritism would become more evident after WWII in the Arsenal's technical documents associated with the Korean War. This favoritism gave that form prestige outside of the Arsenal too.

Street (1984) recognizes that literacy tends to be defined by particular insti-tutions associated with economic and political clout. He also asserts that it is erroneous to favor one form of literacy over others, because different literacy skills may be valuable in different contexts. The New London Group (1996) echoes this valuing of multiple modes of representation and the multiple liter-acies associated with them. Brandt (2001) observes that the same organizations identified by Street (1984) as influential in defining literacy also tend to encourage, facilitate, and benefit from literacy in certain ways; she calls this "sponsorship of literacy."

In the theoretical framework of the New Literacy Studies and sponsorship of literacy, I provide a detailed case study of how technical communication practices at a specific workplace illustrate the value of using a variety of multiple modes of representation to communicate and the value of esteeming all modes, recognizing the importance of developing multiple literacies. While I apply existing theory about multimodality and multiliteracies, I illustrate how ineffective multimodal rhetoric reflected in the accident contributes to our understanding of technical communication and multimodal rhetoric today. I also introduce the phenomenon of literacy sponsorship to multimodal studies. Such discussion helps to link scholarship in multimodal design and rhetoric with technical communication through specific workplace studies. It also shows the responsibility that profes-sional communication, including both managerial communication and technical communication as a field, has in sponsoring literacy outside of workplaces and encouraging multiliteracies. A literacy conflict occurs when communication practices used in workplaces appear to compete with cultural perceptions of which skills ought to be valued or are discarded as utilitarian within a specific historical context. The multimodal approaches that were developed by govern-ment and industry and used at the Arsenal were discarded after WWII to favor, again, print-linguistic literacy practices. However, they have returned as an esteemed program in the lean manufacturing movement of today. The particular form of training workers at the Arsenal received is espoused today in a number of industries, including manufacturing, finance, and health care. As such, the program acts to sponsor literacy through the technical communication and literacy practices it engages and encourages. Consequently, information about this case is very much relevant today.

COMMUNICATION PRACTICES AND
THE ACCIDENT REPORT

The bomb accident served as a major event in understanding connections between modes of representation used in training and other communication practices at the workplace, so I provide some detail here about that accident and its links with literacy at the workplace as an overview of the book. Considerably more information is provided in another chapter.

In the accident investigation report, Stratton (1943) acknowledges various attributes of practices that depot workers used when unloading and moving bombs into position within storage igloos. Employees received specific training but moved away from some attributes of that training in actual practice. While they were trained to unload 165-pound boxes of bombs using two people, the workers modified it so that only one worker would do the job of two, resulting in dangerous handling of the boxes.

Much of the training at the Arsenal came in the form of visual, aural, and experiential training. Workers were shown how to do a given task through demonstrations and then they practiced doing the task. In some cases, very little time passed between the start of training and the worker assuming work on the task. So new employees were put to work quickly, however the training emphasizing demonstrations and hands-on learning was perceived to quicken the learning process.

Stratton (1943) also notes that a memo received by administrators acknowledged that the fuse assembly of the bombs involved was new and defective in its tolerances. So, not only was a defect that could endanger the lives of others known prior to the accident, but a memo—a print-linguistic document—was written by an officer who was aware of defects in the fuse of the particular bomb. This knowledge was communicated to administrators in a different method than most training was delivered to workers.

While workers received their training mostly through visual, aural, and experiential practices, administrators often communicated through print-linguistic forms of literacy. Workers also read print-linguistic materials, though literacy expectations across these materials and positions varied considerably. Training practices affect how employees perform certain tasks, and these practices include the various ways they are taught about performing those tasks. There are clear differences in literacy practices relative to one's position and relative to the accident's timing. So I discuss findings related to training practices as well as other literacy practices that are involved in the accident, such as writing and reading practices among administrators.

A breakdown in communication occurred somewhere between the original memo and the instruction to the workers to be careful, which could be related to the separation of modes and related literacies practiced by each audience. This breakdown suggests that certain information ought to be conveyed in multiple modes of representation—orally and in writing—to reinforce each other. Indeed, manuals that predate the accident include fewer visuals and cautions than those published after the accident. A relationship between visual, aural, experiential, and print-linguistic literacies exists. Administrators recognized this and used this understanding in addressing the problems that contributed to the accident.

Because it is important to understand contextual factors that affected specific communication practices at the Arsenal, I provide historical information about the period and region in its own chapter. In another chapter, I describe some of the research methods I used, calling attention to some challenges researchers face in historical studies of literacy. The next two chapters provide findings related specifically to the communication and literacy practices that occurred at the Arsenal. First, I break the discussion of workplace practices into sections related to the training. Then I consider other literacy practices that occurred at the Arsenal to understand the sponsorship dynamics associated with them and how they may have contributed to this accident.

After presenting findings associated with these practices at the Arsenal, I present information about practices at school, home, and in the community that may have been affected by those practices at the Arsenal. This information suggests a sponsorship influence within the ecology of the workplace and community. I conclude the book by discussing the implications of such sponsorship in the 21st century. In the remainder of this chapter, I provide theoretical foundations of the study and explain why it is important to study such phenomena.

CROSSING SPHERES OF LITERACY PRACTICES

Shirley Brice Heath (2007) invites researchers to consider the implications that workplace literacy practices have on home and community literacy patterns and vice versa. She states,

> Exploring creatively the need for social connectedness of institutions, such as schools and youth organizations, as well as the workplace, offers us ways to create and tell new stories. As we do so, we have to acknowledge that what may seem limits or losses can be beginnings as well as endings. (p. 376)

Heath and others (see, for example, the collections edited by Hull & Schultz, 2002 and Huot, Strohle, & Bazerman, 2004), including Gee (1996), observe distinctions between literacy practices at home, at school, and in the workplace. Workplace literacy practices focus on task-related activities, such as reading instructions on how to perform a given task, to communicating with a co-worker about a work-related issue or project, and composing reports addressing a work-oriented problem or audience. However, workplace literacy practices are valued within the larger cultural structure.

THE NEW LITERACY STUDIES

Much of the early scholarship in literacy studies suggested that literacy is an autonomous phenomenon—free of bias and manipulation—and that literacy within a culture was the path to economic development and individually to the path toward improved socioeconomic status. However, Brian Street (1984) argues that literacy is best considered within an ideological model that includes the understanding of the meaning of literacy as embedded within the institutions in which it is practiced, including home, school, community, and workplace. Street asserts the following characteristics of literacy within this model:

1. The meaning of literacy depends on the social institution in which it is embedded;
2. Literacy can be known to us in forms that already have political and ideological significance and cannot be separated from that significance;
3. The particular practices of reading and writing that are taught in any context depend upon such aspects of social structure as stratification and the role of educational institutions;
4. The processes whereby people learn reading and writing construct the meaning of it within particular practices; and
5. Referring to "literacies" is more appropriate than to a single "literacy" (p. 8).

Political and economic institutions espouse certain forms of literacy practices, and consequently, these forms become idealized within those settings (Street, 1984, p. 107). The federal government and industry employers comprise such institutions. The ideological position of these forms of literacy, then, is what encourages them to be taught in educational settings. Consequently, technical communication practices have a unique position in forming literacy values.

According to Street (1984) and many other New Literacy Scholars such as Graff (2003), Hull and Schultz (2002), Brandt (2001), and Gee (1996), literacy, then, is understood to be a situated practice that cannot be defined uniformly across the settings of work, home, school, and community. Forms of literacy accepted at home, such as letter writing and diary writing, which includes colloquialisms or vernacular, may be considered too informal to be accepted practice at school or work. The way one reads at home (for leisure) may differ from the way one reads at work (task oriented). However, those practices esteemed most at the workplace become esteemed in education and may be esteemed at home.

Further, Street (1984) espouses the concept of multiple literacies—that it is inaccurate to esteem one literacy over any other—though he recognizes that there are political, economic, and social implications associated with these meanings of literacy. Because of the connections between the political, economic, and social, a metaphor of literacy that has emerged within New Literacy scholarship is that of an ecology of literacy (Barton, 2007; Cooper, 1986). Literacy is affected by the culture and environment in which it is practiced, and it affects that environment as well. Literacy practices are motivated by certain environmental factors, and a culture's practices can affect environmental dynamics.

As a social phenomenon, literacy, in the ideological model, facilitates interaction and meaning-making among people in a variety of settings. This communication and meaning-making potential carries political and economic implications with it. Because institutions carry with them the meaning of literacy, different institutions may espouse particular forms of literacy such that practices that are accepted in one setting may be inappropriate to use in another setting. Consistent with Street's (1984) assertion that literacy is defined by the political and economic institutions within a given culture, Deborah Brandt (1999) has observed that certain entities act as sponsors of literacy, and that sponsorship depends on the political environment and can influence economic dynamics of a given time period.

Brandt acknowledges that, "the history of a period is apprehended through the life span, which sets the material and cultural boundaries within which people live out social and economic relationships with others" (1999, p. 3). Political and economic dynamics of a given period affect how people and institutions make use of literacy practices within that environment. Because of their positions of power and influence within the economic and political environments of a culture, particular institutions such as government, business and industrial entities, and schools can affect literacy practices.

Brandt (1999) describes literacy sponsors as those institutions or entities that encourage, facilitate, regulate, limit, or control literacy development for their own benefit. Such entities use literacy as a form of capital. One example that Brandt presents is that of Dwayne Lowery (pp. 52–57). Before Lowery became involved in union activities, he did not read much beyond newspapers. As he became involved in union activities, though, the union funded a trip for him to Washington, DC, to learn skills needed to be a field representative and to negotiate contracts. Further, when companies shifted to using attorneys to represent them at the negotiation table, the union again provided funding for Lowery to attend a workshop wherein he could learn to read and prepare legal briefs. The union, in this case, acted as a literacy sponsor, supporting Lowery's literacy development so that he could better represent the union's interests when negotiating with companies while also contributing to his own professional development, increasing his value as an asset for the union. As such, the literacy that Lowery learned benefited him, and it became capital for the union. The union benefited by encouraging literacy development and facilitating it with economic support. With the Lowery example, Brandt notes that sponsorship can come in the form of economic support for education and training as well as in the form of encouraging and facilitating a particular practice in the workplace environment. She also demonstrates the evolution of literacy demands relative to the changing nature of bargaining, which moves from discussion with company representatives toward discussions with attorneys representing the company. Literacy standards have evolved over time as the economy and its related technologies change.

In addition to noting Street's (1984) consideration of multiple literacies, rather than a single form of literacy, this dynamic also calls attention to historical issues associated with literacy and the study of it. Both Brandt (2001) and Kress (2003) separately observed multiple literacies evolving across time. For example, wherein a college degree in the 1960s suggested that one had strong print-text reading and writing skills, a college graduate today is expected to have considerable computer-literacy skills. With each generation and time period, certain economic, political, and social factors impact what kinds of literacies major institutions choose to facilitate. Brandt calls such changes "accumulation" of literacy (2001, p. 7). Observing a lack of historical perspective in literacy studies, Graff (2003) and Bazerman (2008) called for studies that examine literacy in historical context, indicating that few such studies exist. Such studies provide a lens that permits researchers to consider economic, political, and social dynamics in hindsight.

Relative to taking the historical perspective toward researching literacy, Brandt (1999) notes that

> of key significance in this approach are similarities and differences in the lives of people who have experienced the same set of structural relations and have lived through the same events. This method is useful for gathering information about changes in the material networks through which people have learned and practiced literacy across time. (p. 3)

An analysis that focuses attention on a particular ecology can inform theories of literacy and contribute to reshaping literacy instruction. It can also close gaps in practices across locations of literacy practices—home, school, community, and work. When such studies focus on sponsorship attributes associated with a given workplace in a specific ecology, they shed light on the role of technical communication in literacy policy.

Workplace Communication, Literacy, and Changing Economies

Hull (1997) examines the relationship between workplace communication practices and literacy. Generally, she observes several connections between literacy and the economy. These include that a growing number of workers are considered to be illiterate as economic and technological changes occur, and illiteracy affects productivity and product quality as people need time to learn how to perform their job correctly and effectively. Throughout the collection, the theme that emerges is one of finding ways to train workers for particular jobs as demand for some jobs weakens and demand for other jobs increases. For example, in Hull's collection, Merrifield (1997) considers the economic impact that layoffs due to a changing market have on communities and the role the government plays in facilitating development through legislation that provides financial assistance to particular groups in high-risk social categories. Merrifield calls attention to the 1982 Job Training Partnership Act that provided funding to displaced workers (p. 277), and Brandt (2001) has alluded to the G.I. Bill, which was used to help returning veterans fund their pursuits into higher education and professional training (p. 87).

Graff (2003), Heath (2007), Hull (1997), and Brandt (2001) note themes associated with historical influences that affect literacy practices. Such themes are part of the literacy ecology; certain economic and political dynamics that are part of a given time period influence the literacy ecology. The field of literacy studies includes research that considers ecological dynamics of literacy within social environments, but it lacks close examinations of historical treatments and implications of certain forms of sponsorship, especially related to technical communication. Heath (2007) makes some observations in the Epilogue about the political consequences of her study. However, these are limited to observations of how government policies did not support the changes she implemented in the school on the basis of her study. Also, the observations were made 10 years after her study was published. Contextual dynamics are affected temporally, and more studies of factors that contribute to a given temporal ecology of literacy are needed. Further, Graff (1979) asserts that any definition of literacy must be able to account for cultural and temporal differences in literacy dynamics.

The Influence of Workplace Literacies Across Other Spheres of Literacy

Home, school, community, and workplace environments interact to influence a given ecology of literacy practices. Heath (1993) considers, among various sociolinguistic dynamics within three communities in a particular locale, how the literacy practices at a textile mill relate to the literacy practices at home and in school. Generally, she finds that there is a considerable disconnect between the practices at work and those espoused in school. Teachers in Heath's study attempted to encourage students to learn Standard Written English by inviting workers from the mill to talk to students about the reading and writing practices at the workplace, hoping that as the workers talked about those practices, students would become motivated to learn reading and writing skills. However, the teachers came to understand that "there were few occasions for writing or reading extended prose in these jobs . . . they usually filled out forms or summarized orally to someone else who 'wrote up' the 'final' report" (p. 311). As the children did not see how the literacy they were taught in school would help them get a job, this disconnection sabotaged literacy learning efforts at school and home. The forms of literacy that were valued within the technical communication practices at the mill indicated to the children what literacies were valued in that society.

While some scholarship, such as Heath's (1983) study, considers schools and community programs as literacy sponsors, few studies exist that are related to the examination of a particular workplace as a literacy sponsor. Studies that examine workplace literacy practices, like Winsor (2000), have considered relations between workplace practices and certain community-related literacy practices. Also, C. S. Johnson (2009) examines historical practices at a single company and changes therein, and J. M. Staggers (2006) provides a systematic analysis of the relationship between certain communication practices within a given workplace and their affect on a particular geographic community in her award-winning dissertation about risk communication and the development of the atomic bomb. Evidence from the study I conducted shows the power that literacies valued in a workplace have outside that workplace. While a variety of literacy practices were used, the one that was valued most in the historical context is obvious, while others are discarded.

Multiliteracies/Multimodality and Technical Communication

Technical communication, especially, integrates multiple modes of representation and engages multiple literacies. Many recent studies of literacy practices have discussed multiple literacies and multimodality (Gee, 2003; Huot et al., 2004; Mayer, 2005; Mishra & Sharma, 2005; Moreno & Mayer, 2000; New London Group, 1996). The New London Group (1996) identified various modes of representation, including the print-linguistic, visual, aural, experiential, spatial, and gestural; any combination of these is considered to be "multimodal." While it is generally inappropriate to esteem one literacy or modality over any other, the New London Group as well as Kress and van Leeuwen (2001) observe that there may be contexts in which certain modes "are more powerful than other modes" (New London Group, 1996, p. 63). For example, Moreno and Mayer (2000) found that multimodal instruction that integrates narration and images (aural and visual modes) facilitates better learning than using either mode individually.

Because of the context-bound nature of literacy practices, it is impossible to value one form of literacy over any other generally; however, as cultural and economic dynamics change over time, certain literacies emerge as more valued than others in particular contexts. Brandt (2005) suggests that literacy may be considered a form of capital, observing that literacy opportunities are "configured . . . in rationales of production and profit-making" (p. 194). Studies in technical communication report various kinds of literacies needed to be able to function in particular professional settings as professional specialization continues (Cicourel, 1981; Haas, 1994; Northey, 1990; Olsen, 1993). Likewise, Scribner and Cole (1981), Heath (1993), and Street (1984) were among the first to observe multiple forms of literacies at work in a given culture in and beyond the workplace and contexts in which different literacies worked in those cultures. Further, Northey (1990) observed different kinds of writing and related literacy practices practiced at different levels in accounting firms, suggesting a literacy hierarchy within organizational structures.

As Graff (2003) notes in the "Introduction to Historical Studies of Literacy," historical studies of literacy in context may illustrate repercussions of certain practices. He asserts that, "Social attributes . . . and historical contexts, which are shaped by time and place, mediate literacy's impacts, for example, on chances for social or geographic mobility" (p. 125). Much literacy research, in fact, describes literacy using the ecology metaphor; that is, the context in which literacy occurs needs to be considered within the analysis of literacy itself. Cooper (1986) explains that writers, as part of an ecological system, interact with each other to form systems (p. 368). She argues that characteristics of an individual piece of writing determine and is "determined by the characteristics of all the other writers and writings in the systems" (p. 368). The writing that one does in one context (e.g., the workplace) interacts with writing one does in other contexts (e.g., school), and the writing others do in those contexts affects one's writing as well.

At home, children learn to communicate with others with a close social bond. School literacy practices tend to emphasize learning how to write Standard Written English and developing basic reading skills ranging from poetry to longer pieces of literature. Several studies since Heath's (1993) study and call have documented relationships between home and school literacy practices (see Hull & Schultz, 2002, for example). Generally, these studies suggest the need for a closer link between home literacy practices and those practiced in educational settings; the more literacy practices are linked across settings and assimilate home literacy practices, which are the most frequently practiced and with which people are most familiar, the better learning and application of practices become.

Phelps (1991) acknowledges a geography of knowledge relative to writing She describes that the knowledge product of a given local community may be evidenced in documents immediately located in that community as found in particular local case studies (p. 876). As Heath (1993) found, when there are disconnections in literacy practices across contexts, the result may be one of confusion, affecting literacy practice and learning. Further, Brandt (2005) continues to observe the esteem given to print-linguistic literacies in workplaces, suggesting a link between certain literacies and economic potential generally. However, studies document the use of visual literacies in the workplace as well (Brumberger, 2008). The study I report in this book shows the material importance of connecting practices across levels within an organization as well as across contexts of practices. The particular workplace environment and ecology beyond the workplace environment affect what literacies are valued.

By considering the original context in which certain workplace literacy practices occurred and studying the relationship between those practices and practices within the community, literacy scholars can understand a specific case in which employers adjusted literacy expectations to workers' skills and benefits associated with that. Such a historical study can also uncover any ideologies at work within that ecology and consequences associated with this adjustment; this study attempts to ascertain those consequences in addition to understanding sponsorship dynamics within the political and economic environment of the time period.

PURPOSE AND SCOPE OF STUDY

Studying a particular temporal and geographic ecology of literacy practices and sponsorship can shed light on ecological factors that affect a workplace's influence as a literacy sponsor and implications associated with that sponsorship. The historical dynamics at the time created the particular employer and caused a large migration to the area. Because of the anticipated migration, the federal government constructed a housing development that included a school. Further, because of the literacy traditions and limited skills of the employees, the employer used a certain program to train employees. These particular attributes—the creation of the institutions, the migration, and the literacy backgrounds of the employees—create an ecology in which certain literacies are sponsored. Also, employers can make an effort to accommodate literacy levels in the workplace.

Specifically, the study reported here considers the literacy ecology involved when the government built an arsenal near Fieldview, Ohio (names are pseudonyms), which was a farming community and to which many people from various literacy backgrounds, including Appalachians and African Americans, migrated for work during World War II. The sites were significant locations during WWII, and the effects of the Arsenal and various dynamics were felt beyond the end of the Korean War.

More specifically, the primary research question associated with this study is, As a major employer in Fieldview Ohio, how did the government and operators of the Boomtown Arsenal sponsor literacy for its employees and for the community of Fieldview from 1940 to 1960, and what ecological factors influenced this sponsorship? I limit the study to this period because by 1960, operations at the Arsenal had declined dramatically, and children of migrants who stayed in the area had finished high school. Issues related to this question that contribute to a definitive response include ascertaining particular practices and literacies at the workplace and how they were used and valued; what evidence of sponsorship of literacy there is in terms of literacy institutions outside the workplace; and what practices occurred in the community, at school, and at home that reflect similar valuing of literacy practices at the workplace. Findings from the study carry implications for current technical communication literacy practices and sponsorship thereof.

Specifically, this analysis examines the relationship between modes of representation and exigencies involved in the particular historical context. As indicated previously, literacy practices include print-linguistic forms of literacy as well as other modes of representation used for communication purposes, including the visual, aural, experiential, spatial, gestural, and combinations thereof. Also, as I acknowledged earlier, the New London Group (1996) and Moreno and Mayer (2000) argue that certain modes of representation and certain combinations of modes may be more powerful for certain purposes and in certain contexts. The fields of technical communication and business communication have evolved to value the modes and literacies identified by the New London Group.

Much research in workplace communication and literacy studies uses ethnographic approaches, in which the researcher observes and possibly participates in the literacy activities of those studied (Szwed, 1981). However, Krippendorff (2004) observes that content analysis is the best approach to research in the absence of actual observations of physical activities. Because this is a historical study, I rely on content analysis using interviews and document analyses.

I interviewed those who lived the historical literacy experience, however I also reviewed documents from the particular workplace involved to better understand what kinds of literacy practices actually occurred there. I use the interviews to understand practices outside of the workplace as well as inside it. These interviews represent the participants' recollections of their practices, and the interview questions are open-ended, so people could share their stories freely. Furthermore, reviewing documents from the workplace provides a more comprehensive picture of literacy practices there. Case study designs integrate the local contextual considerations. More information about methodological dynamics is presented in Chapter 2.

The Goal of the Book

This is a historical case study of the communication practices at a particular workplace and the literacy relationship between the Arsenal and Fieldview, Ohio, during a certain time frame. In this book, I explicitly articulate the relationship between technical and managerial communication practices and their role in literacy sponsorship within that historical context; but I also link it with current communication and literacy sponsorship practices. This book, then, contributes an understanding of the social and material consequences of certain forms of sponsorship associated with workplace communication. This study can contribute to the fields of technical communication, managerial communication, and literacy studies in no fewer than three ways:

1. Offer insight into understanding what contributes to effective multimodal rhetoric in technical communication;
2. Consider potential implications of a balancing act pertaining to multiple modes of representation and their related literacies associated with certain forms of literacy sponsorship; and
3. Identify implications associated with literacy sponsorship and multimodality relevant to today's economy, educational philosophies, and visual culture.

More than a book about technical and managerial communication in a workplace, the analysis presented in this book offers insight into the interrelationships of workplace, home, and community literacy activities in terms of the influence that literacy sponsors, especially those in business and industry, have on that ecology. Sponsors must react to certain environmental dynamics. Sponsors may accommodate certain literacy practices to their benefit. However, decisions that sponsors make about valuing certain literacies over others and how they attempt to negotiate practices they value with those less valued carry serious implications for sponsors as well as those affected by the sponsorship. While Brandt's study (2001) emphasized print-linguistic literacies because of their importance in the 20th century, I consider various modes of representation identified by the New London Group (1996) and sponsorship of those literacies. As such, I expand on Brandt's conception of literacy sponsorship, and I discuss connections between sponsorship and the ecology of literacy metaphor.

There is a rhetoric embedded within literacy practices, and this study suggests potential dangers when disconnections between literacies and related modes of representation occur. While the Arsenal is no longer operational as it once was, given Heath's observation (1996) that "what may seem limits or losses can be beginnings as well as ends," a historical study can inform new directions in workplace and technical communication studies and literacy learning. An examination of the literacy dynamics at work within the particular ecology of a given historical period can provide a better understanding of how practices intersect locations of home, school, community, and work; and operating and learning efficiencies associated with these intersections may occur.

The study has implications in the current economic environment in which workers need to relocate to different geographic areas and transition to different or new industries, adjusting to new discourses while learning to perform their work. Similarly, literacy demands change in a dynamic technological environment in which virtual game systems that assimilate behavioral and physical/gestural modes of representation are used to facilitate training and other workplace practices. Workplaces also want employees to be able to transition quickly into their job without needing much training. Multimodal forms of training facilitate learning in such environments.

The major sponsorship elements studied are the government and the contractors who operated the Arsenal. Literacy policies are developed and implemented at the federal level and include input from business and industry, as implied by the makeup of the Secretary of Education's Commission on the Future of Higher Education (2006). This study, consequently, can inform future policy development. There are reports of an economic crisis in which major local employers are closing, affecting hundreds of jobs and local economics (Mackinnon & Towell, 2009). In this same economy, certain kinds of employers are having difficulty filling positions; jobs are available, but people lack certain skills necessary to be hired for them (Schleis, 2009). Such reports signal, once again, that literacy demands are shaped by political and economic dynamics and that literacy is truly context-dependent. As economic demands shift the kind of work that people do, literacy demands change as well.

Technical communication, along with managerial communication practices, is a large part of this sponsorship of literacy. The study reported here, then, is an effort to advance a theory that recognizes the intersections that affect a given ecology of literacy and ways that institutions that act as literacy sponsors affect that ecology.

CHAPTER 2

Methodology and Issues in Historical Research

The raison d'etre of content analysis is the lack of direct observational evidence.

—Krippendorf, 2004, p. 39

As I described in Chapter 1, literacy practices are part of a particular, dynamic ecology in which people's uses of literacy are affected by environmental variables and may affect those environmental variables. Similarly, Fleckenstein, Spinuzzi, Rickly, and Papper (2008) assert that, "An ecological orientation to research emphasizes the need for research diversity: multiple sites of immersion, multiple perspectives, and multiple methodologies within a particular discipline and research project" (p. 401). Such research considers a variety of contexts within a given study, including workplaces, churches, neighborhoods, and historical dynamics (p. 401). Examination of multiple sources of information can triangulate data to ensure reliable findings and conclusions.

Krippendorff's (2004) words preface the chapter because the principal analytical method of the historical study that constitutes a majority of this book is content analysis within case study design. Much of the theoretical grounding for my research methods comes from content analysis theorists Krippendorff, as well as Silverman (2006), Crano and Brewer (2002), Weber (1990), and Holsti (1969) because they consider the general dynamics of such analyses. However, because of the role of multimodal design and rhetoric within the technical communication dynamics at the Arsenal, it is also important to consider work by multimodal content analysis theorists Bateman, Delin, and Henschel (2007), Mathiessen (2007), Rowley-Joliet (2004), Royce (2007), and Unsworth (2007) relative to document analyses involving multimodal attributes.

Yin (1984) acknowledges six "sources of evidence" within case study design, and three of them—documentation, archival records, and interviews—are a part of this study. I apply content analyses to the data collected from these sources. This approach is necessitated largely because of the constraints of a historical study of literacy practices: In a historical study, one cannot observe practices in use as many ethnographic studies do; one must analyze content from sources available to facilitate study of those practices.

Historical research in workplace communication and literacy practices presents its own challenges. In this chapter, I detail the methods I used in the historical case study. Throughout, I also articulate efforts to address particular issues in historical studies of multimodal literacy practices in workplaces; specifically, the following six issues emerged:

1. Concerns about the memory of older adults who were interviewed,
2. Interview sampling relative to representation of the population,
3. Dealing with sensitive workplace information,
4. Coding multimodal documents,
5. Accounting for missing pages in archived documents, and
6. Ascertaining actual use of printed materials.

Graff (2003) and Bazerman (2008) encourage historical studies into literacy practices. By sharing details of research methods and means to address challenges that arise, researchers can develop sound methods to facilitate such studies.

FOUNDATION OF THE STUDY

Though the Arsenal's munitions operations have been closed and the space is used primarily as a training site for ROTC programs, its presence in the general community still maintains a considerable place in the history of the area. Periodic news stories recall its activities during WWII and the Korean War; and given that so much toxic material was used in the production of munitions during its history, stories about decontamination efforts appear in the news as well. Because of the size and significance of this unique site, I began to look into its relationship with the literacy practices of people who lived near it.

The primary people involved in my efforts to pursue the study were a colleague whom I knew was connected with the county historical society, a friend of hers who was involved in Fieldview's Historical Society, and the archivist at the Arsenal. The woman who was involved in the Fieldview Historical Society gave me access to some print materials that described general historical information about the community, the school district, and the Arsenal.

At this time, I also contacted the Arsenal about my interest in studying literacy dynamics there, and I eventually was referred to its archivist. I came to learn that she had been there for 2 years organizing the various documents at the Arsenal as part of an effort to develop a database of archival records. She gave me access to historical summaries, which were produced at irregular intervals during the Arsenal's operations. I reviewed these for information about the Arsenal's history and literacy practices during the periods summarized. I did not review all of the historical summaries; rather, I focused on those that were produced during the

WWII years, the Korean War years, and the Vietnam War years because the Arsenal ended manufacturing operations permanently and storage operations were dramatically reduced after the Vietnam War. The Arsenal operated primarily as a Reserve Officers training ground after the Vietnam War.

In this review of documents I focused on literacy-related themes identified by Brandt (2001), such as identifying what kinds of materials workers read or wrote, how training occurred, and the kinds of backgrounds workers came from as well as historical information to help me understand certain contextual factors that influenced these practices. This review helped to frame the research questions for the empirical study as well as help me to think about some coding attributes to consider as perhaps a first round of coding. I elaborate on the development of my coding scheme later in the chapter.

In addition to this data, I also reviewed several scholarly works describing historical patterns in migration of workers, educational background, and industry in the state and nationally (Hobbs, 1998; S. A. Johnson, 2006; Nelson, 1995; Rodabaugh, 1975; Walsh, 1995). The remainder of this chapter describes the methods used for the empirical study of literacy practices within the particular ecology and the issues that arose as I pursued the study.

MIXED-METHODS DESIGN

This study includes content analyses of interviews and archived documents. A majority of it relies on qualitative analyses, however it also includes some quantitative analysis relative to readability dynamics. A special issue of College Composition and Communication (September 2012) dealt with issues in archival research. A few of the articles in that issue raised questions about what counts as archives and ethics of archival research. Gaillet (2012) raises several questions concerning archival research, including how to locate archives and ways to triangulate findings (p. 38). Many workplaces do not like for industry secrets to be reported in research studies, however once a government record is declassified it becomes open to the public for consideration. One way to locate new archives is to review entities that the government once owned or operated to ascertain any declassified materials. Also, my effort to use interviews and archival records is considered appropriate for triangulation.

Yin (1984) characterizes case study research as an empirical inquiry that investigates a contemporary phenomenon within its real-life context wherein the boundaries between phenomena and context are unclear and in which multiple sources of evidence are used (p. 23). MacNealy (1999) acknowledges that "case studies tend to rely heavily on interviews" (p. 203). Yin (1984) further acknowledges that, "one of the most important sources of case study information is the interview" (p. 88). Generally, interviews allow researchers to conduct open discussions about phenomena under study with participants.

Yin also states that, "documentary evidence is likely to be relevant to every case study topic" (1984, p. 85). In particular, he explains the usefulness of reviewing documents to triangulate information from other sources as well as to enhance that information (p. 86). Yin cautions that researchers who use documents and archival records must try to understand the contexts in which the materials were developed and for what audiences (p. 88). So interviews can help ascertain those contexts.

Silverman (2006) points out that analysis of documents also facilitates an understanding of choices writers made regarding how to represent information for particular audiences (p. 152). Consequently, my coding includes study of patterns in the modes used to represent information in documents as well as reports of other modes of representation used to communicate information at the Arsenal. As a historical case study, I rely on analyses of documents that have been archived at the Arsenal as well as on data collected from interviews with members of the Fieldview community, some of whom worked at the Arsenal.

INTERVIEW METHODS

Brandt (2001) explains that lifestory research integrates "historical, socio-logical, psychological and phenomenological inquiry . . . [and includes] struc-tured and less structured interviews" (p. 10). In this historical case study, I use a similar approach to Brandt's: interviewing those who lived the historical literate experience. Some of the interview participants worked at the Arsenal during the period of study, while other participants who did not work at the Arsenal lived in the community, attending the school district and participating in community organizations. These interviews reflect the participants' recollections of their practices. A concern about these interviews is the sampling/population associated with it.

RESEARCH ISSUE 1:
HISTORICAL STUDIES—
MEMORY AND ETHICAL RESEARCH

Many of the interview participants were over 80 years old at the time of the interview. Much of the literature on concerns about memory related to inter-views with elderly participants identifies issues associated with disorientation and dementia (e.g., Eeles & Rockwood, 2008; Taub, 1980). These studies recommend tests of cognition and recall prior to beginning formal interviewing. None of my participants had indications of disorientation or dementia, and people with whom I spoke about the participants acknowledged belief that the participants had a "sharp" memory.

Kirkevold and Bergland (2007) and Decker and Adamek (2004) acknowledge memory deficits as a concern in qualitative interviews with elderly participants. Both recommend interviewers take additional time to establish rapport with their participants and to use a less-structured interview, which Decker and Adamek acknowledge can yield "richer" and "more comprehensive" data than that gathered from structured interviews (p. 61). This could affect the accuracy of data. I attempted to address this concern by triangulating data sources; if a number of participants identified the same dynamics or experiences and practices, then I could conclude that the information was reliable.

RESEARCH ISSUE 2: INTERVIEW SAMPLING

The total population of possible participants is limited to people who lived in the particular community during that time frame and those who may or may not have worked at the Arsenal too. Consequently, my sample includes some who worked at the Arsenal and some who did not work there but lived in the community at some point during the period of study. However, because this is a historical study, and one must consider that people move to other places and people may pass away, my sampling is limited to those who stayed in the area and were still living. This affects sampling relative to race and gender representation of the total population.

In attempting to facilitate a sampling that provided a cross-section of the population of the area during the time period, I used two approaches to recruit participants. I made announcements about my research project, inviting people to participate, at two of the local historical society meetings—one at a November meeting attended by approximately 100 people, and the second at the February meeting, attended by approximately 30 people. At each meeting, a speaker was scheduled to present information about local history. Prior to the presentation, society leaders permitted me to announce my project and facilitate volunteers. Generally, I introduced myself as a graduate student at a local institution who was studying the relationship the Arsenal had with the community, and I was focusing on reading and writing practices of people who worked at the Arsenal and/or lived in the community between 1940 and 1960. These announcements produced a list of 23 names.

I called anyone who wrote their name and contact information in response to these announcements or were referred to me. In each of these conversations, I acknowledged the nature of my study, what their participation would entail (an interview lasting approximately 30 minutes), and asked if they would be willing to participate. Most (18) of these people were willing to participate, and I made an appointment to interview them at their home, with IRB approval of the research protocol.

Race/Ethnicity

Because the sample is derived from announcements at a social gathering and referrals, this is a convenience sampling. Also, all of the volunteers from this initial set of announcements and calls were Caucasian. According to Rodabaugh

(1975), African Americans comprised approximately 2.5% of total war-related employment in 1940 and 8.2% of total employment in war-related industries in 1945. African Americans also worked at the Arsenal and lived in the community, however, while records documenting the total number of employees and their gender and ethnicity breakdowns were maintained, almost all of these records were lost or destroyed. The only record of this data that I could find was within the historical summary of July to December of 1943. According to this summary, African Americans (termed "negroes" in the archived documents) comprised at most about 11% of the workforce there during that period (see Table 2.1).

In an effort to include the experiences of African Americans in the interview sampling, I asked the president of the Historical Society if she knew of any African Americans from the community who may be willing to participate. She gave me a list of a dozen African Americans whom she knew as a student in the school district, not knowing where they currently lived.

I searched for their contact information in the phone book and online, and I contacted those six for whom I could ascertain contact information. Three of these were via phone, and three were via email. In spite of these efforts to recruit African Americans, none volunteered. Consequently, my sample includes only Caucasians.

Gender

The sampling also includes more than twice as many females as males (13 females, 5 males). Women tend to outlive men by approximately 5 years: the average life expectancy is 80.5 years for White females and 75.3 years for White males (Shrestha, 2006). According to the 1943 historical summary of operations at the Arsenal, until September/October of 1942, males made up a large majority of the workforce at the Arsenal. However, in September and October of 1942, given the shrinking labor market, "experiments" were conducted to

Table 2.1. Gender and Race Breakdown of Employees

Month	Male	Female	White	"Negro"	Total
July 1943	3947	2622	6289	280	6569
Aug	5026	2855	7366	515	7881
Sept	4799	2775	7065	509	7574
Oct	4643	2803	6854	592	7446
Nov	4503	3105	6731	877	7608
Dec	4359	2941	6473	827	7300

Source: From Atlas Power Company, History of BOP, Vol. II, p. 8.

ascertain which production operations could be performed by women. "These tests demonstrated the fact that many operations heretofore thought too strenuous for female labor could be performed efficiently by them" (p. 313). So gender-related employment shifts, especially on load lines, where most of the assembling of munitions occurred. Tables 2.2 and 2.3 provide an illustration of gender-related employment patterns on two load lines, showing this shift.

While female employment eventually exceeds male employment on Load Line I, this was not the trend at the other three load lines reported. This data is

Table 2.2. Load Line I Employment
(1941–1943)

Date	Male	Female	Total
8/25/41	34	0	34
9/30/41	174	11	185
10/25/41	398	6	404
11/30/41	684	15	699
12/30/41	630	69	699
1/31/42	975	60	1035
2/28/42	1420	60	1480
3/30/41	1916	60	1976
4/30/42	1826	60	1886
5/30/42	1767	60	1827
6/30/42	1834	60	1894
7/30/42	1664	60	1724
8/30/42	1489	60	1549
9/30/42	1346	60	1406
10/31/42	1268	125	1393
11/30/42	1111	578	1689
12/31/42	868	726	1594
1/30/43	670	707	1377
2/27/43	624	617	1241
3/27/43	446	422	868
4/30/43	537	614	1151
5/31/43	524	867	1391
6/30/43	408	643	1051

Source: From History of BOP, Vol. I, p. 314.

Table 2.3. Load Line II Employment
(1941–1943)

Month	Male	Female	Total
Nov. 1941	61	0	61
Dec. 1941	200	0	200
Jan. 1942	291	0	291
Feb	370	0	370
March	530	0	530
April	257	0	257
May	459	0	459
June	508	0	508
July	599	0	599
Aug.	422	0	422
Sept.	426	0	426
Oct.	367	0	367
Nov.	551	0	551
Dec.	671	179	850
Jan. 1943	646	283	929
Feb.	612	349	961
March	436	470	1006
April	382	192	574
May	233	112	345
June	305	246	551

Source: From History of BOP, Vol. I, pp. 321-322.

important because it shows the gender-related inconsistencies in employment across different time periods included in the study, which can affect the degree to which the interview sampling represents the population of employees who worked at the Arsenal. The interview sampling includes two females who worked at the Arsenal. Also, the interview questions invite participants to share information about practices that any of their relatives who worked at the Arsenal shared with them in an effort to include more of the experiences of those who worked at the Arsenal. One of the interviewees who did not work at the Arsenal spoke of an aunt who worked there. Consequently, the interview data includes perspectives from female workers.

Table 2.4 shows a breakdown of the interview participants.

A total of 11 of the 18 interviewed were students in the Fieldview school district at different points during the period studied. Five worked at the Arsenal during the period studied and two worked in the community; one of these was a teacher in the Fieldview school district. Six (33.3%) of the participants were native to Fieldview prior to 1940, and five (27.8%) migrated to the area between 1940 and 1950. Six migrated to the area between 1950 and 1955, and one migrated to the area after 1955. Of the 12 who migrated, 5 migrated from within Ohio, while 4 migrated from more than one state away. Nine of the twelve who migrated to the area did so for work-related reasons. Four migrated for work specifically at the Arsenal. Eleven participants reported that they, their parent(s), or a relative worked at the Arsenal. One participant was unable to recall what kind of work their relative did. Table 2.5 shows the breakdown of the 10 Arsenal work positions reported.

Generally, then, the interview sample includes representation of people who were native to the area prior to the Arsenal's construction as well as those who migrated into the area for work at the Arsenal, those who worked at the Arsenal— male and female—and who held different positions there, and those who did not work at the Arsenal but experienced community, home, and school-related

Table 2.4. Interview Participants

Gender	Student?	Arsenal employment	Native/ Migrant	Migrated from?
Female: 13 Male: 5	11 attended the school district	5 participants worked there 6 indicated that a relative worked there	Native: 6 Migrated: 12	Ohio: 5 More than one state away: 4

Table 2.5. Interview Participants by Position at Arsenal

Position	Number of participants
Construction/carpentry	3
Secretarial	2
Line/labor	3
Supervisor	2

literacy practices. However, a limitation of the study is that it does not include representation from African Americans.

Interview Questions

The interviews are semistructured to open-ended (Silverman, 2006). The script that I used for the interviews is in Appendix A. Most of the questions follow those Brandt used in her study (2001). While she asked her participants about various reading and writing practices, Brandt's questions were open-ended to allow participants to provide responses that included narratives of particular practices. I used similar questions because of the open-ended nature of the questions and because they allowed participants to speak freely of their experiences.

Generally, the script acted to guide the questioning, however, after ascertaining demographic information, the interviews tended to occur as conversations with the participants responding at length to several questions. These responses often integrated information that could answer other questions in the set because of relationships between questions. For example, question 17 asks, "How much reading was required of the training program?" and question 19 asks, "How much reading was required in the job(s)? What kind?" Some participants included job-related reading requirements as they responded to question 17. Consequently, most of the interviews did not follow the script as it appears. Sometimes, a response facilitated a question not in the script as a follow-up to clarify or explain some information.

Heath (1993) acknowledges a concern about how the way she spoke to interview participants may have affected their responses (pp. 264–265). Considering this concern, I tried to be careful with my speech and even dress so as to make the participant feel comfortable. I wore casual, sporty clothes—generally a dress shirt or pull-over shirt along with casual pants. I wanted to make the participants feel at ease while also appearing as a professional researcher. I tended to sit between 3 and 8 feet from the participant, depending on the floor plan of the room in which the interview occurred. Usually we sat at a kitchen or dining room table, while a few interviews occurred in the participant's living room, with the participant sitting at one chair and me at another chair and an end table or coffee table between us.

With the consent of the participants, I recorded the interviews with an Olympus DS-2 digital voice recorder. Because the voice recorder is able to record meetings without an external microphone, I placed the recorder in a location approximately equidistant between the participant and me. Prior to beginning the recording, I gave the participant a minute or two to look over the script to help them think about some responses ahead of time. Then I asked the participant if he or she was ready to begin and began recording upon ascertaining their readiness. After the response to the last question, I thanked the person for participating and then stopped recording. In some cases our conversation about the interview resulted in their sharing some more information, which I wrote in a notebook that I had and in which I noted certain experiences or practices that seemed to stand out.

Transcription

After recording the interview, I transcribed each using the voice recognition software Dragon NaturallySpeaking 12.0. As I played back the recording, I spoke the words into the software, which then generated a transcript. As I did this, I often stopped the recording and went back to ascertain the accuracy of the transcription, typing in corrections where necessary. Also, after generating a complete transcription, I reviewed the transcription with the recording, making any corrections as needed.

Because transcription should document phenomena relevant to the research questions (Geisler, 2004; MacNealy, 1999), I did not transcribe pauses or inflections or other oral language phenomena. My focus was on the themes of literacy practices articulated in the response, not how the response was said. While phonological attributes may affect meaning, the participants all spoke the same dialect of English as the researcher. Furthermore, pauses document a break in language flow; they do not document reasons for that break. The researcher could only speculate as to the reason for a pause: was the person thinking about a response, was the person trying to clarify facts in their mind as they recalled them prior to articulation, was the person distracted by someone or something in the room at the time of the pause? On more than one occasion, for example, a pet entered the room where the interview occurred, distracting both the respondent and interviewer momentarily.

Because I did not videotape the interviews, I also could not integrate non-verbal cues or gestures that participants used into transcription. Consequently, I did not document places in which such gestures may have affected an oral response from me. An example of this is the following exchange, during which Roger responded to my question about training he received:

R: You pick up a shell, put it in a vice

Me: And they showed this to you or you didn't have anything to read but they showed you how to do this?

R: No, no. It was very . . .

Me: Hands-on?

R: Hands-on. Very hands-on

Roger explained the process of drilling a shell casing. As he did this during the interview, I recall that he used his hands to gesture the motions involved. He had said nothing about reading a manual or receiving classroom instruction for the procedure, so I asked him how he came to learn that process. He began a response and paused to think of a term that described how he learned the process.

During that pause, he gestured with his hands. The exchange prior to this pause and his gesturing prompt my offering of the term "hands-on."

These items can be considered limitations of the methodology. Had I video-taped the interviews, I could have documented more precisely a number of gestures and expressions not articulated. Also, by not documenting pauses or inflection attributes, I limited the data documented to oral language. However, in both cases, the limitation did not dramatically affect the data collected relative to the scope of the study.

Unit of Analysis

Holsti (1969) identifies six generally used units of analysis for content analysis: Word or symbol, theme, character, sentence or paragraph or item (pp. 116–117). For this study, I used "theme" as my unit of analysis, which Holsti labels as the most useful unit for content analysis (p. 116). Holsti characterizes the "theme" unit as an assertion made about some subject. In using "theme," the researcher must break down a particular statement into a single theme or themes to facilitate tabulation into categories. I identify categories later, however understanding theme as my primary unit for content analysis helps explain transcription and document selection approaches.

Categories and Coding

In this section, I explain coding of responses to these questions. Most questions are coded relative to the presence of certain practices relative to those themes identified above; for example, "home writing" is coded relative to the presence of "journals and diaries," "correspondence" and "work reading" is coded relative to "correspondence," "reports," and "manuals." These are coded as nominal data (categorical data) rather than as interval data (indicating some quantitative value) to avoid suggesting any value of literacy level or skill associated with any particular practice. Nominal variables do not suggest any particular ordering, ranking, or value associated with them. If, for example, I were to format "home writing" as an interval or ratio variable and give "correspondence" a higher value than "diaries," it would suggest that skills associated with correspondence are valued more or are more rigorous than skills used for writing diaries. I list all categories and related codes in Appendix B.

In coding passages from transcripts, I considered the information a particular question sought and then looked for certain codes represented in the response. I did not count how many times a given code was acknowledged, only whether it was acknowledged. An example is from the exchange below:

> Me: Did you do any reading or writing outside of school? You mentioned the Methodist Church writing that you did. Anything at home or leisure reading newspapers?

Steve: I did letters—did a lot of letter writing and those types of things . . . to grandparents and I did some reading but not a lot of reading. I was a worker—had a paper route and paper routes and did a lot of work through school. We didn't have a whole lot of money. We just did those types of things.

This exchange would result in my coding this participant's home writing activity as limited to "letter writing." I also coded for non-print-linguistic literate practices any participants identified, because I considered the New London Group's (1996) conception of multiliteracies and multiple modes of representation within the scope of the study. A few of the participants who worked at the Arsenal acknowledged the hands-on training associated with their work there, which I coded as visual, aural, and experiential (combining spatial and gestural modes identified by the New London Group).

An example of coding a relative's workplace literacy experience, from the same interview, is below:

Me: Okay, you mentioned that your father worked at the arsenal and that he participated in the construction of it. What did he do at the arsenal when he started there?
Steve: He was in supervision and supervision in various titles for number of years and he retired in 1982 after 42 years at Boomtown arsenal.

For this exchange, I would code the work experience represented as the person's "parent," the position as "supervisor," and employment years as "over 40." If the participant shared information about the relative's experiences at the Arsenal, I included that information in the dataset, which would then provide a more comprehensive picture of literacy practices at the Arsenal.

Analysis

In addition to analyzing data for each category, I also reviewed data regarding general trends across categories, such as the kind of reading and writing activities that occurred at school and at home to see if any thematic patterns arose. For example, if participants acknowledged more print-linguistic forms of reading and writing (such as reading newspapers at home and writing essays for school) than they acknowledged of visual practices (like the map-drawing acknowledgment related to school), it suggested an emphasis on print-linguistic skills over other literacy skills across spheres as a theme.

RESEARCH ISSUE 3:
WORKPLACE STUDIES AND SENSITIVE INFORMATION

Because of the Arsenal's position in national defense and security policies associated with it, people who worked there may not have felt free to share literacy narratives about the workplace practices. There was a security policy that forbade workers from talking about their work there. So I needed to be able to ascertain sensitive information that interviewees may not have been willing to offer. I was able to address this through reviewing archived documents that had been declassified.

Not only can review of workplace documents triangulate interview data, it can also provide information that interview participants may not share. Interview data sheds light on recollections of general practices of each person; review of actual workplace documents helps to understand specific practices and expectations of writers and readers there and to triangulate interview data pertaining to the workplace practices.

Another issue connected to this point is that companies tend not to want industry secrets reported in research studies. However, the workplace I studied was a government-owned, contractor-operated site, which means that documents are available once declassified. Because they were produced during a time of war and concerned production innovations, they had been classified and hidden from researchers until released.

Document Analysis Methods

I reviewed documents from the Arsenal to better understand what practices actually occurred. As I mentioned above relative to triangulating with interview data, reviewing as many different types of documents from the workplace provided another necessary data source to build a comprehensive picture of literacy practices there. I include in this corpus historical reports/summaries that operators of the Arsenal prepared at irregular intervals, depending on how active the site was; standard operating manuals (SOPs); newsletters; incident reports; and building specifications. The total corpus included in this study numbers over 40 documents. In the next section, I detail data sources within the context of data collection. Review of the variety of documents sheds light on specific literacy practices across different levels of the organization.

I reviewed several documents at the worksite in order to address the research question: What literacy practices occurred at the Arsenal? Krippendorff (2004) observes that researchers can determine an effective sampling size by reviewing a large corpus of the entire population and selecting a smaller corpus of texts that represent patterns across the larger corpus (p. 123). I explain my process later in this section.

The archivist at the Arsenal gave me access to an electronic database listing archived documents. However, she also acknowledged that this list was incomplete—many documents had not yet been included in the system since she had been working on it for some 3 years at that point and had more to review. Documents not yet in the system were still disorganized, and part of her job was to bring some organization to the Arsenal's materials.

I studied 43 archival documents from the Arsenal to

1. ascertain relationships between different modes used for representation and literacy expectations associated with them to address the research question regarding "What relationship between print-linguistic literacies and visual literacies existed at the workplace?" and
2. ascertain literacy expectations through readability testing to address the research question, "What literacy practices were required in the workplace?"

I drew randomly from the listing of available documents based on the number of documents archived relative to different types of documents. That is, first I ascertained some parameters of the number of documents of a given type and differences in those numbers (e.g., manual/SOP versus routine reports), and I did not use any systematic means of drawing texts for the sample other than skimming the catalog and picking from those published during the period of study. Table 2.6 shows a breakdown of the types of documents and the number of each included in the sampling.

This distribution represents to some degree the distribution of available documents relative to each type. That is, there are more newsletters than any other kind of document, so newsletters make up the largest proportion of documents in the sample, and the format of these ranges considerably across time (this is explained more in Chapter 5). There also seemed to be several routine reports, including historical summaries, but the format and style of these tended to be consistent. Specifically, these tended to be 8½" × 11", standard print format, integrating headings and subheadings and 1" margins, and using 12-point font size throughout. Also, the same kinds of sections were included and ordered similarly across documents. There were also several manuals/SOPs, but the format and style of them varied over time. In particular, a 1945 bomb SOP used

Table 2.6. Documents Sampled

Type of document	Number sampled
Manual/Standard Operating Procedure	6
Newsletters	16
Routine reports	11
Special reports	5
Building specifications	5

landscape format for graphics and text, and it integrated photographs associated with each step with the textual instructions for each step. Several SOPs from the 1950s used portrait format for text and photographs, and the photographs are appendixed relative to the textual instructions. There were fewer special reports, such as incident reports, however all of these were formatted alike. Also, there were several building specification documents, and these tended to follow the same format and style.

Sampling of Documents

Holsti (1969) acknowledges that sampling documents for content analyses generally occurs in a two- or three-stage process: identifying a list of sources, drawing a sampling of entire documents, and sampling a limited number of pages from within documents (p. 130). This study uses all three of these stages for the analysis of archived documents.

Sources

The first round of coding, as described above, helped me to identify potential sources and characteristics of those sources. I reviewed those in the database to ascertain the number of documents and particular types of documents available. From this review, I was also able to ascertain that I would need to use a stratified sampling. Crano and Brewer (2002) and Krippendorf (2004) acknowledge that stratified sampling involves recognizing unique subgroups within a given population. In the Arsenal's population of documents, there were different types of documents available—newsletters, routine reports, special reports, and manuals. These different kinds of documents represented different subgroups of documents; each written for a unique subgroup of the worker population. Subgroups are identified in Table 2.7.

Because different readers and writers were associated with each subgroup and my research questions sought to ascertain reading practices across different kinds of employees within the Arsenal, I needed to distinguish between these subgroups.

Table 2.7. Employee Literacy-Related Subgroups

Document	Subgroup
Newsletters	General work audience
Routine reports	Upper management/administration
Special reports	Upper management/middle management
Manuals	Line workers and supervisors

General Sampling

I then sampled approximately 25% to 33% of each type of document available and listed in the database. While scholarship generally does not identify a specific percentage of a given population as ideal for sampling, I used this figure because it would offer a reasonable amount of the population of each type of document from which general observations about the documents could be made. Generally, scholarship calls attention to challenges in sampling for content analysis, but the researcher's responsibility is to try to ascertain a way to generate a representative sampling of the population. The population size was unknown, however finite, and drawing on a relatively large percentage of the known items would provide a reasonable sampling for the entire population. I recorded aggregate information, including the number of print-linguistic text pages and the number of pages that included graphics and the type of each kind of graphic to facilitate analyses of proportions of each. My research questions included analysis of relationships between different modes of representation and related literacy demands, so recording this information would help to address those questions. There was quite a bit of variability in formats and lengths of the various documents. This sample size would give me a good idea of the range of modes used within each type of document while representing a relatively large sampling of the population. The first round of coding helped me to identify particular categories and coding for this analysis, as indicated above.

Selective Sampling

Finally, I sampled a limited number of pages to facilitate readability testing. Generally, I randomly opened the document by placing my first finger at a random location (not counted or predetermined) on the side of the closed document and flipped to that page. I then transcribed a few articles (newsletters) or complete page of text (other documents) and pasted that transcription into the testing tool, which is described later in this chapter. I did the same thing in flipping to another page. I repeated this to collect several samplings from a given document.

Further, I tried to sample documents representing the time period's range; that is, I sampled documents of each type (except for building specifications, all of which were from 1940 or 1941) from the 1940s, 1950s, and 1960. Just as including interview participants with experiences just beyond the time period range allows for analysis of any changes in policy, such coverage with documents allows for analysis of policy trends in the workplace. As I acknowledged above, some documents varied in format and style across time periods, so it was important to sample a range of documents within a given time period as well as across time periods associated with the study.

In analyzing documents, I tabulated data relative to aggregate information. Aggregate information considers phenomena associated with an entire document, and these variables are provided below. This information would help in answering questions pertaining to certain literacy practices for particular kinds of documents and their readership. In the next section, I describe the development of specific categories I used for the aggregate analysis.

Rhetorical Purpose of Document and Related Readerships

Generally, there appeared to be several differences in the use of graphics across different kinds of documents relative to their general readership and literacy skills associated with those readers. However, these differences could also be attributed to the purpose of the document. Much scholarship finds that certain kinds of graphics are more useful than others for particular rhetorical purposes: generally, for example, photographs closely approximate physical objects they represent, making them useful in instructions, while tables effectively represent numeric data for reports in which such information is presented (Gurak & Lannon, 2007; Helmers, 2006; Kolin, 2009; J. Murray, 2009). Also, graphics placed within the text of a document are considered primary graphics (immediately relevant to the purpose of the document), while those placed in appendices are generally considered secondary in nature, supplementing the text information (Markel, 2010, p. 141; Oliu, Brusaw, & Alred, 2010, p. 69).

Further, as acknowledged in Chapter 1, Mayer (2001) and others (see, for example, the collection edited by Mayer, 2005) support the use of multiple modes in education and training materials because the different modalities can reinforce each other, or certain modes may appeal more to certain readers than to others, and including both will facilitate learning for different kinds of learners. J. Murray (2009) also finds the value of redundancy in providing print-linguistic text alongside of related images in multimodal compositions (p. 32).

Unit of Analysis

As with the interview analysis, the unit of analysis for the documents was that of themes. Certain themes pertaining to patterns in the use of the various modes of representation used in various documents and related literate expectations/ demands emerged from the review of documents. The categories I identify below attempt to facilitate empirical analysis of those themes.

Categories: Approaches

Development of categories to facilitate analysis is debated in the literature, especially regarding use of a grounded-theory approach, as indicated above (Crano & Brewer, 2002; Holsti, 1969; Krippendorf, 2004). Generally, one can use an *a priori* approach identifying categories prior to the actual collection of data; or one can use a grounded-theory approach, letting categories that facilitate analysis emerge as one collects data. Strauss (1987) acknowledges that such analytical categories can be identified by the researcher at any of several phases in the research. Krippendorff (2004) acknowledges five different ways to define the unit(s) of study for content analysis: physical distinctions, syntactical distinctions, categorical distinctions, propositional distinctions, and thematic distinctions (pp. 103–109). Two that work for the study of texts that include print-linguistic and visual modes of representation—multimodal representation—are physical distinctions and categorical distinctions. Because this study examines such relationships, I used similar categories. Categories related to physical attributes of a message that the literature typically identifies, though, include amount of space on a page devoted to the object(s) of study, size of the object being studied on a given page, and frequency of occurrences of a given attribute (Crano & Brewer, 2002; Krippendorff, 2004; Royce, 2007; Matthiessen, 2007). Krippendorff (2004) also acknowledges that frequency of occurrences can characterize categorical distinctions—the frequency that a given category of a variable occurs suggests something about its value. While different documents may use different relationships relative to their purpose, examination of a single kind of document or modes of representation across different readerships relative to a single purpose may find particular trends.

While little debate about *a priori* categories exists, Holsti (1969) and Krippendorff (2004) encourage a grounded-theory approach, while Crano and Brewer (2002) discourage it. Crano and Brewer acknowledge that content analysis is a form of observational research, which necessitates identification of units prior to making observations (p. 247). Consequently, as acknowledged previously, I use mostly *a priori* categories. However, Holsti (1969) encourages a grounded-theory approach, acknowledging that the categories must reflect the research questions and that the standardization of a set of categories assumes a large corpus of research on a given phenomenon (pp. 101–102). Further, Krippendorff (2004) suggests that a grounded-theory approach is appropriate when he states that units "emerge in processes of reading and thus implicate the experiences of the analyst as a competent reader" (p. 98). As noted above, Strauss (1987) explains that attributes of a grounded-theory approach can occur at each step in the research process (pp. 25–32). I used it principally within the concept-indicator phase, which Strauss acknowledges directs coding of certain empirical indicators (p. 25). Indicators are data that are indicators of a given "concept the analyst derives from them" (p. 25). Strauss goes on to explain that through comparing similarities and differences across indicators/data, a category emerges and related codes can be refined accordingly (p. 25). While most of the categories I used have been identified to some degree in literature previously acknowledged, I observed two important attributes of the documents and used a grounded-theory approach toward accounting for those items; consequently, my coding reflects certain issues within historical studies not presented in the literature. Specifically, placement of certain graphics and the presence of any missing pages became important to document. These are two more issues I needed to address.

RESEARCH ISSUE 4:
CATEGORIES FOR WORKPLACE MULTIMODAL PRINT MATERIALS

The categories I used differ somewhat from those generally identified in the literature, because the literature on multimodal pages tends to deal with content analyses of newspapers/magazines, which tend to use column space as a measure of use. Fewer studies have analyzed workplace reports or manuals. In addition to several commonly used categories, two additional categories that emerged pertinent to aggregate data were that of location of visuals relative to the text that describes them: "Co-location" (or Relative Location) and "missing items."

The more recent work in multimodal analysis seems to examine relationships between text and moving images/animation on a computer screen (e.g., Wysocki, 2001) or comparing text-based representations to animated representations of the same content (e.g., Unsworth, 2007). Further, Unsworth acknowledges that certain kinds of visual representations may articulate explicit semiotic relations between the visual and the viewer or the writer and reader (p. 332). For example, a photograph showing actual people who represent the viewer will be more meaningful to that viewer than a diagram showing some representation of the viewer (p. 348). So I coded for photographs as well as drawings or diagrams, because a photograph may carry more rhetorical meaning for a reader in a certain context—a manual, for example. Use of a lot of photographs compared to any other kind of graphic may suggest a rhetorical decision by the writer associated with the reader's literate background.

I also coded these graphic and textual distinctions because Matthiessen (2007) acknowledges that combinations of graphics and text have been "a feature of literate cultures," but that the division of labor between images and text has undergone "significant" changes over time (p. 29). An emphasis on diagrams and images and photographs suggests an emphasis on visual literacy skills as a theme, while an emphasis on print-linguistic text suggests an emphasis on print-linguistic literacies as a theme.

Bazerman (2008) asserts that "site-specific questions must attend to the particular character, opportunities, and difficulties of gathering data at the site as well as to the kind of analysis the data will allow" (p. 306). Consequently, my data collection methods attempted to address specific site-related issues, like missing pages in documents, lack of color photography, and age/memory-related issues associated with interview participants. While Brandt (2001) does not examine any textual materials, Heath (1993), Street (1984), and Scribner and Cole (1981) consider texts within their design to understand relationships between literacy practices in different settings. Generally, for example, Heath (1993) codes such documents relative to the kind of verbal and visual representations associated with them, though she does so very generally, and she uses narrative analysis. For example, she observes that "reading material in the mill, beyond section names and signs marking restrooms, lunchroom, trash cans, soft drink machines, etc., is limited to information on the bulletin boards" (p. 234). She does not report an empirical textual analysis showing frequencies of certain relationships or types of text or graphics. However, I attempt to code these variables for empirical study here to facilitate mixed-methods analyses because of potential relationships they may reveal between literacy expectations of readers and the material presented in the documents. This analysis is consistent with the consideration of multiple modes of representation that is associated with this study. Extensive use of visual representations of information in documents suggests an emphasis on visual literacy skills, while extensive use of print-linguistic text in documents suggests an emphasis on print-linguistic literacy skills.

Categories Used

In this section, I identify specific categories and coding procedures for the document analyses. I also show examples of that coding. Certain kinds of categories tend to be identified across content analysis theories. Because of their relevance to this study, Table 2.8 shows the categories I used to code data relative to aggregate materials (entire documents).

Table 2.8. Document-Coding Categories

Category	Scale of measure: Nominal, ordinal, interval, or scale	Codes
Type of document	Nominal	Manual/SOP Routine report Special report Newsletter Building specification
Total number of pages	Scale	
Pages with only text	Scale	
Pages with only graphics	Scale	
Combination pages (pages with both text and graphics)	Scale	
Co-location	Nominal	In-text Appendixed Some in-text, some appendixed

Figure 2.1 shows an example of a page that would be coded as "text-only"; note that it includes a list presented in alphanumeric print-linguistic text. I included such listings as text. While lists are generally included as a form of graphic representation, I did not code lists largely because I include

MONORAIL SYSTEMS

INSTRUCTIONS TO BIDDER:-

In addition to a lump sum price for two (2) monorail systems described in these specifications, the bidder must quote separately upon each monorail system complete with its specified number of trolleys.
The bidder is also requested to quote a delivered price and delivery date for additional trolleys, if ordered at the same time as the complete monorail system, as follows:

A price for 20, for 60, or for 100 additional trolleys for Carriers 27-P.

A price for 200, for 400, or for 600 additional trolleys for Carriers 28-P.

A price for 10, for 20, or for 30 additional trolleys for Carriers 53-P.

The bidder shall submit with his bid, drawings and descriptive literature in sufficient detail to thoroughly show the essential and safety features of the complete installation which he proposes to furnish. Failure to furnish such information will be considered cause for rejection of bid.
The bidder must state the shortest time after the awarding of contract within which he can guarantee to fabricate, deliver and erect, ready for operation, the complete monorail systems with minimum number of trolleys, and also the shortest time for delivery of the additional trolleys.
The bidder shall state (and be prepared to demonstrate) the horizontal effort or force necessary to start a fully loaded carrier in motion upon a straight level rail, and the force required to continue it in motion, at a uniform speed. This information will be a major consideration in the awarding of the contract.
In the Receiving and Painting Building, WP-3, roof trusses are spaced 20'-0" center to center and the monorail tracks must support continuous lines of fully loaded carriers. If additional structural supports (in addition to those shown on the drawings) are required to properly support the monorail system in this building, or at any other heavily loaded point or points throughout the Loading Line, the bidder should call attention to the fact that such supports are not included in his bid, and at the same time shall supply full information as to the supports required and the loads which they must carry. The Contractor is requested to quote a separate price for furnishing and erecting these additional structural supports.

1

Figure 2.1. Example of text-only page.
Source: From Specifications for Monorail Systems for Melt Load Line No. 1; Specifications no. 40, p. 1. Document courtesy of the U.S. Army.

SOPs and manuals in this study, and these documents revolve around lists that may be broken down into further lists. For example, an SOP for making a bomb includes a step-by-step listing of tasks, which are then broken down into a listing of steps to perform that task. An example of an image-only page is in Figure 2.2.

I coded such organizational charts and other graphics that integrated print-linguistic text with visually formatted representations as graphics instead of as hybrid text/image because, while the text is part of the graphic representation, the visual representation in which the text appears is emphasized on the page and graphic. However, I coded templated forms, which one would complete by filling in blanks placed alongside prescribed print-linguistic text, as print-linguistic text.

While the entire page in Figure 2.2 is occupied by a graphic, the feature that would include it as image-only is that it is the only representation on the page.

An example of a combination page is in Figure 2.3.

The page includes standard print-linguistic text as well as photographs, which are discrete from each other. To code the page provided in Figure 2.3, I first identified it as a combination page, then acknowledged the number of graphics on the page (2), what percentage of the total page is occupied by both graphics (approximately 33%), indicated the size of each graphic (the graphic to the left

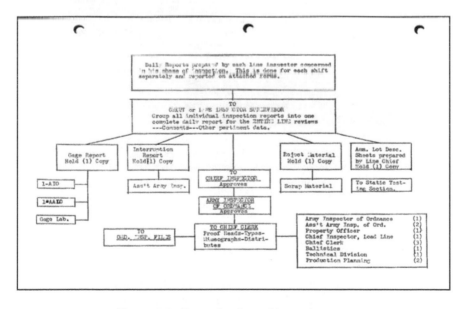

Figure 2.2. Example of graphics-only page.
Source: From Ordnance Inspection Manual, p. 170.
Photo courtesy of the U.S. Army.

Figure 2.3. Example of combination page.
Source: From Newsletter, May 1942, p. 4.
Document courtesy of the U.S. Army.

would be coded as "20%–33%" of the total page, while the graphic on the right is coded as less than 20% of the total page). Finally, I listed the kind of graphic each was (photograph).

While the abovementioned scale categories are identified in content analysis literature, I observed other phenomena for which I had to develop categories and codes. Strauss (1987) explains that categories should reflect the phenomena being analyzed, consequently, categories may emerge from the data within a particular study.

Location Category

It became clear that visuals in manuals and SOPs produced after 1950 were positioned after the text information, while they were placed within the related textual information frequently in such documents produced before 1950. I included the relative location code because of the various placements that I observed. Bateman et al. (2007) consider such positioning of information within their category of "rhetorical structure": "how the content is divided into . . . main material, supporting material" (p. 155). Generally, appendixed information is considered supplemental material, while graphics placed within the text are considered primary to the purpose of the document (Markel, 2010; Oliu et al., 2010). Because of this temporal difference in locating graphics and the potential association with perceptions of the relationship between the text and graphics, I coded for these different locations.

RESEARCH ISSUE 5:
ACCOUNTING FOR MISSING PAGES

Some pages were missing from some documents, having been destroyed or lost, and some of these pages contained graphics, according to references in pages that exist. Some of these were textual pages, while others were appendixed pages. For example, some documents included textual references to graphics found in appendices, and the appendices were missing.

Rather than omit these documents entirely from analysis, I attempted to include the fact that some pages were missing in the analysis while including the materials that did exist. Excluding the entire document from the analysis and including only documents that were complete omitted a portion of the population of documents. I tried to identify frequency of certain phenomena, such as the number of text-only and graphics-only and combination pages as well as types of graphics used. A given document that was missing pages may have had more of certain types of pages than I could have observed, but I could only analyze the material I observed. So coding for missing pages lets the reader understand that there may have been more graphics or text-only pages or more of a certain type of graphic present, but that I could not document it. In any case, I tried to minimize the number of missing pages included in the data so as to minimize the effect those missing pages had on the analysis.

Readability Tests

Finally, readability tests are part of content analyses (Crano & Brewer, 2004, p. 262; Holsti, 1969, p. 89; Krippendorff, 2002, p. 58) because they are a measure of the reading skill expected of a given audience. Because one of the related research questions in this study is to understand literacy requirements of employees who worked at the Arsenal relative to how their background may have affected literacy practices, I included analysis of literacy levels expected of and practiced by employees as evidenced in readability tests of various documents. To facilitate this analysis, I randomly sampled three to four passages from a selection of each type of document—manuals/SOP, routine report, special report, newsletter. That is, I turned groups of pages in no systematic fashion, coming to a page that may have had graphics on it or not and selecting a passage from that page. I drew this sampling in this way to establish a set that represented readability attributes for the entire document.

I retyped two or three entire articles (newsletters) or passages (manuals, reports, building specifications) of an average of over 200 words per passage, with a range of 70 words to 422 words. I then applied five different readability tests, all of which measure the grade level necessary to understand the content using different algorithms to each sampled passage. To run the test, I used the website http://www.online-utility.org/english/readability_test_and_improve.jsp. This website allows the user to copy and paste text passages into a box, and the program will calculate results for several different readability tests.

The site facilitates several tests, including Gunning-Fog, Coleman-Liau, Flesch-Kincaid, ARI, and SMOG. All of these are associated with measures of readability relative to a particular grade level. That is, the resulting output number associated with any of these tests reflects the grade level needed for a reader to be able to understand the passage. However, each uses a different set of variables to arrive at the grade output.

Both Krippendorff (2004) and Crano and Brewer (2002) acknowledge that the Flesch-Kincaid test is the most-used test, acknowledging that even the U.S. Department of Defense uses it to measure readability of its documents; consequently, I report only data related to the Flesch-Kincaid test in this study. This test, which is derived from a test Flesch developed in 1943, considers the average number of words used per sentence and the average number of syllables per word. Examination of the mean grade-level readability score associated with the various documents in the workplace helps to identify patterns of reading expectations of targeted audiences of those documents. If the mean of a particular kind of document is lower than for another document, this suggests that the audience of the first kind of document is expected to have a lower grade-level reading skill than readers of the second document.

RESEARCH ISSUE 6:
ASCERTAINING ACTUAL USE OF DOCUMENTS

While I conducted content analyses of several documents, a problem that occurred to me was that just because documents exist in a workplace, employees may not actually read them, or they may use them a certain way not stated within the documents themselves. A researcher may find information regarding a document's intended audience and purpose within the document or they may be able to infer by nature of information in it. However, how it was read is difficult to ascertain with just the documents.

To address this problem, I asked people whom I interviewed and who worked at the site to what degree and how people used the actual documents (source triangulation). Included in my interview sampling were 11 people who worked at the site or whose relatives worked there. According to Krippendorff (2004), interviews allow participants to recall their interaction with various documents and identify which were of primary importance for them and how they interacted with any workplace documents associated with various tasks.

Qualitative Analyses

I used qualitative analytical methods in this study. Yin (1984) encourages analysts to identify a particular theme or unit on which to facilitate analysis of case study data. He generally discourages use of statistical analyses within case study design—single case study and multiple case study—largely because variables studied are unlikely to have any "variance" (p. 113) and selection of particular cases is "not based on any sampling logic" (p. 124). However, he encourages using a "pattern-matching" approach to analysis to facilitate internal validity in a single case study and replication of a given study to other cases, thereby enhancing external validity. Data from different observations within a case study may show a certain pattern from which findings and conclusions are drawn.

Traditionally, analyses of content tend to focus on the presence of given phenomena being studied and patterns that emerge. Within this study, I analyzed patterns and trends of certain phenomena. Further, literature that describes analyses of visual information tends to call attention to certain categories of attributes of graphic data relative to salience. These include use of color, relative size, richness, and sharpness (Kress & Van Leeuwen, 2006; Rowley-Joliet, 2004; Tufte, 2006a, b; van Leeuwen, 2003). Consequently, my coding included relative size and the type of graphic (e.g., photograph versus diagram). Color photography and printing were not available for part of the period under study, so I did not include them as a characteristic of study. Bateman et al. (2007) consider this dynamic in their category "production constraints," which may be attributed to availability of color photography and/or macro-economic concerns about costs of using such (p. 155). Further, because I examined documents published over a period of 20 years, I also observed changes that occurred in their content and format over this period. Several documents published in a given 2- or 3-year time frame exhibited similar attributes, but the attributes differed with documents published more than 5 years apart. I discuss these progressive changes over time as shifts in attributes and general trends.

CONCLUSION

In this chapter, I have identified details associated with my research methods for this study, including identifying specific issues associated with historical research into multimodal workplace practices and how I tried to address them. This study uses interviews and document analyses within case study research design.

The sample for the document analyses is a good representation of the range of literacy practices at the Arsenal during the period of the study. Counting frequency of occurrence, testing readability, and measuring means of the categories involved allowed for an understanding of literacy practices and expectations at the Arsenal. Also, interviews with several members of the community, including former employees of the Arsenal, helped to triangulate data associated with the interviews as well as data from archived documents.

APPENDIX A:
Interview Questions

1. Were you born in the Windham area or at what age did you move to the area?
2. School(s) attended?
3. College? Degrees attained: what; when completed?
4. Age when you began working in area?

General Adult Education

5. Did you attend any adult education/training classes/workshops?
6. When?
7. Where?

Adult Education—transition-community-sponsored training

8. What training programs do you recall were available to help with you or your family's transition to life in Northeast Ohio?
9. What community-based training programs were available to develop reading and/or writing skills?
10. What reading skills were developed in this program?
11. What writing skills were developed in this program?
12. How did the skills developed in this program help you or your relative(s) at home, education, and/or in any work you or they have done?

Job Search

13. Did you or anyone in your family work at the Arsenal? What position(s)?
14. Did anyone help you complete job applications?
15. What questions related to reading/writing skills were asked in any job interviews?

Work-Related Skills

16. What was the Arsenal's training program like?
17. How much reading was required of the training program?
18. How much writing was required of the training program?
19. How much reading was required in the job(s)? What kind?
20. How much writing was required in the job(s)? What kind?

Home/Other Work Literacy

21. How did the skills you/your relatives learned in this program help at home or in any other job you/they held?
22. What other reading and writing training did you/they receive?
23. How did this training help you/them at home or at any other job you/they held?

APPENDIX B
Interview Categories and Codes

Category	Scale (nominal, interval, ratio/scale	Response Codes
Native	Nominal	Yes, No; arrived before 1940 No; arrived between 1940 and 1945, No; arrived between 1946 and 1950 No; arrived between 1951 and 1955 No; arrived between 1956 and 1960
Migrate from	Nominal	Not applicable Within state One state away More than one state away Outside U.S.
Migrate reason	Nominal	Not applicable Affordable housing Work, self; not arsenal Work, self; arsenal Work, parents; not arsenal Work, parents; arsenal
Grade entered	Scale	

Grade completed	Scale	
Graduation year	Scale	
College	Nominal	No Yes; some college Yes; completed associates degree Yes; completed Bachelor degree Yes; completed Graduate degree
Arsenal work	Nominal	No Yes; reporting my own experience Yes; reporting parents' experience Yes; reporting siblings experience
Years employed	Scale	
Position	Nominal	Line worker/laborer/inspector Supervisor, manager Administrator
School writing	Nominal	None Poems Essays Plays Poems/essays Poems/plays All
Work writing	Nominal	None Manuals Correspondence Reports Correspondence and manuals Correspondence and reports Manuals and reports All
Home writing	Nominal	Notes/reminders Diaries/journals Letters Diaries/journals and letters All

Others home writing	Nominal	Did not write
		Parents/aunt/uncle worked Arsenal;
		   wrote letters
		Spouse, sibling worked Arsenal;
		   wrote letters
		Parents/aunt/uncle worked Arsenal;
		   wrote diaries/journals
		Spouse, sibling worked Arsenal;
		   wrote diaries/journals

Home read self	Nominal	Did not read Newspapers Magazines Books Newspapers/magazines Newspapers/books All
Home read to	Nominal	None Letters to parents Letters to children Stories to children Newspapers/magazines to children

Others read home	Nominal	None
		Parent/aunt/uncle worked Arsenal;
		   read letters
		Spouse/sibling worked arsenal
		   read letters
		Parent/aunt/uncle worked Arsenal;
		   read newspapers/magazines
		Spouse/sibling worked Arsenal;
		   read newspapers/magazines
		Parent/aunt/uncle worked Arsenal;
		   read books

Work talk at home	Nominal	Not applicable No Yes
Community write	Nominal	Newsletter Promotional material Reports

Community read	Nominal	Newsletter Promotional material Reports
Community programs	Nominal	None Library-sponsored Church-sponsored Civic-sponsored
Influences	Nominal	None identified Parents Teacher Other community members Parents and teacher Parents and other community members Teachers and other community members All
Library recall	Nominal	No experience conveyed School-related only Included leisure Worked there
Adult literacy program	Nominal	None Yes; basic reading/writing skills Yes; career-related program

CHAPTER 3

Historical Context

In this chapter, I describe historical attributes that provide a social and political context for the study. This chapter provides information specific to the region associated with this study as well as about national dynamics that affect regional dynamics. Such information is critical to understanding the ecology in which the practices associated with the study occurred. Generally, the national economy at the time of WWII was shifting from a farm economy to war industry, and there was considerable migration to particular areas where war-related plants existed. Migrants came from particular kinds of communities, and these migrations affected not only community populations but also made for an economically and educationally diverse population in those communities.

On December 29, 1940, President Franklin D. Roosevelt delivered one of his Fireside Chats via radio, and this particular speech has become known as "The Great Arsenal of Democracy" speech (Appendix C). In it, he calls for national focus on restructuring the American economy and workforce to aid Great Britain and allies in their fight against Nazi Germany, recognizing the global threat against democracy the war already going on in Europe posed. Roosevelt referred to the construction and development of arsenals throughout the country. The Boomtown Arsenal was among the arsenals built under the program discussed in this speech; indeed, it was already under construction at the time of the speech. However, the speech provides considerable information to help understand the political context of the period.

THE GREAT ARSENAL OF DEMOCRACY

An awareness of Roosevelt's speech helps us to understand the ecology in which the government and operators of the Arsenal acted between 1940 and 1960 and certain themes of the literacy ecology of Fieldview. This speech articulates much of the political, economic, and ideological attributes that influenced the literacy dynamics from 1940 to 1945 and beyond. It also announced the reason for the construction of the Boomtown Arsenal (in addition to others) as well as the grounds for development of the Fieldview community, where many employees would live. Review of this speech identifies particular historically contextual attributes that suggest certain ideologies in addition to the political environment within this particular ecology of literacy.

Early in the speech, Roosevelt called attention to the uniqueness of the period relative to a global threat to democracy and the demands it placed on a unified goal among the population. He stated,

> The Nazi masters of Germany have made it clear that they intend not only to dominate all life and thought in their own country, but also to enslave the whole of Europe, and then to use the resources of Europe to dominate the rest of the world.

These words illustrate the threat that democracy faced at the time and the sense of urgency for immediate action.

Further, he called for a national, unified effort to help the Allies even before the United States was officially engaged in war and the implications for the future. He acknowledged,

> The people of Europe who are defending themselves do not ask us to do their fighting. They ask us for the implements of war, the planes, the tanks, the guns, the freighters which will enable them to fight for their liberty and for our security.

By reinforcing the threat to the United States and calling on industry to assist the Allies, Roosevelt engaged the country's resources in the war effort, even though the United States was not officially involved.

Subsequently, he explicitly acknowledged what the populous, including commercial and industrial interests, must do to react to them, explicitly challenging the notion that the United States would become a full participant in the war. Indeed, he alerted the populous to the degree of their participation and industrial work ahead:

> Nine days ago I announced the setting up of a more effective organization to direct our gigantic efforts to increase the production of munitions. The appropriation of vast sums of money and a well-coordinated executive direction of our defense efforts are not in themselves enough. Guns, planes, ships and many other things have to be built in the factories and the arsenals of America. They have to be produced by workers and managers and engineers with the aid of machines which in turn have to be built by hundreds of thousands of workers throughout the land. In this great work there has been splendid cooperation between the government and industry and labor. And I am very thankful. . . .
>
> I want to make it clear that it is the purpose of the nation to build now with all possible speed every machine, every arsenal, every factory that we need to manufacture our defense material. We have the men, the skill, the wealth, and above all, the will.

Given the nature of the effort, he alluded to the United States as "the great arsenal of democracy."

This speech articulates the political and economic environments prevalent just before WWII that prompted the construction of the Arsenal. Roosevelt espoused the political ideology of quick production of arms in favor of the present global demands and against future economic and political considerations. He also attempted to minimize resistance that might occur in hiring people to work in the munitions plants in an effort to maximize potential production.

FIELDVIEW

The government began purchasing the land on which the Arsenal would be constructed in August of 1940, and construction began the next month—3 months before Roosevelt's speech. Operations on the first manufacturing line at the Arsenal began in August of 1941—only 4 months before the Japanese attack on Pearl Harbor. The second manufacturing line began operations just before that attack (late November), and the third manufacturing line was operating just after it (December 10). By May of 1942, nine manufacturing lines were operating, and the storage depot area had been established (Pfingsten, 2010). Because of the location of the Arsenal and its housing, Fieldview experienced the largest population growth of any locale in the country between 1940 and 1950; growing from a small farming village of approximately 300 residents to a town of over 3,000—approximately 1100%. The Arsenal displaced over 200 family farms. The federal government invested much financial support in Fieldview's infrastructure. It financed the construction of the village's fire station, a school, and library in addition to housing. Further, the demographic makeup of the community changed around 1960, by which time the government-supported housing development was sold to a private company and converted to "Section 8 housing." This development was built as the Arsenal was being constructed to facilitate housing for those who would move to the area for employment at the Arsenal, and it made up a large portion of the community. During this 20-year period, many people who lived in this development were either employees of the Arsenal or veterans who returned from WWII or the Korean War and were pursuing higher education opportunities at nearby institutions with help from the G.I. Bill.

Indeed, during the height of WWII, the Arsenal employed over 12,000 people, which fell to under 3,000 shortly after the war. The Arsenal never employed more than 3,000 during the Korean War, and during peacetime, it employed a few hundred personnel, mostly in storage operations.

The Arsenal is located almost entirely within Parker County, with a small section extending into in Tree County. Boomtown, the county seat of Parker County, is located about 5 miles southwest of the Arsenal. Boomtown is the largest city in the immediate area of the Arsenal, and it is where workers would cash their paychecks and where most would socialize. Fieldview is a small community located to the immediate north of the Arsenal, and it experienced the most dramatic population and economic change of any locale in the country at the time of WWII.

The labor pool in the immediate area fell short of demand for workers there, so recruiting efforts occurred in several neighboring states and in the South. It is unclear specifically how many workers migrated to the area for work at the Arsenal, but archives document that African Americans and people from Appalachia as well as the South migrated there. Appalachians were no strangers to the area however.

The first migration of Appalachians into the area was in the 1910s as migrants moved from Appalachian farming areas to farms in the Midwest. Wages were higher and the land less difficult to negotiate. Kentuckians tended to move to Cincinnati and Columbus, while West Virginians moved to Akron and Cleveland (Barr-Capper, 2006; Hobbs, 1998). S. A. Johnson (2006) notes that, "more newcomers became homeowners and considered Akron their permanent home" (p. 1). So large was the migration of West Virginians to northern Ohio that Carl Feathers (1998) notes metaphorically that, "by the early 1920s Akron was the capital of West Virginia" (p. 23). So a social network for additional migrations had been established before WWII. Between 1940 and 1970, states Hobbs (1998), about one million people moved from the mountains to Ohio; most were White, but about 50,000 were African Americans (p. xiii).

As with the Appalachian migrants, African Americans migrated to the North in the early 1920s and through the 1930s. African Americans had worked mostly in meat packing plants, steel mills, and foundries, and they found similar work in the North. Rodaburgh (1975) notes that in 1940, African Americans accounted for 2.5% of the total employment in war-related industries, and by 1945, they accounted for 8.2% of total employment in war-related industries. Between the labor shortage and an executive order by Roosevelt, many industries relaxed discriminatory hiring practices.

According to Ben Wattenburg (2006), Executive Order 8802 prohibited discrimination in defense and government jobs, contributing to an influx of African Americans into the war plants in the North and West. This also contributed to the gradual development of a Black working class (Rodabaugh, 1975). Furthermore, African Americans now moved to places in the country where they could vote, which they could not do in any appreciable numbers in the segregated pre–World War II South.

Finally, Wattenburg notes that

> the war's impact on blacks is reflected in the numbers. Between 1940 and 1950, the black population of Mississippi went down by 8 percent. The black population of Michigan went up by 112 percent. In 1939, black males earned 41 percent of what white males earned. In 1947, they earned 54 percent of what white males earned.

REGIONAL MIGRATION DURING WWII

The Midwest experienced considerable in-migration during WWII, partly because of the new importance that the farms in the region had in providing food and to the war-related industries in the region. Nelson (1995) acknowledges that Michigan, Ohio, Indiana, and Illinois had a net gain of 600,000 interstate migrants. Many came for farmwork that was offered throughout Ohio, but many also came for the work in war-related industries in Ohio.

WWII AND THE BOOMTOWN ARSENAL

As I have indicated previously (Remley, 2009),

> As the U.S.'s involvement in World War II began, the government looked for locations where it could establish plants that would produce ammunition and weapons. For security reasons, many of these locations were small cities that were away from the seaboard and experienced various weather patterns that would make radar tracking difficult for enemies. (p. 98)

Boomtown, Ohio, was one such location. The Arsenal was developed as two facilities—a manufacturing plant that would be operated by private industry and a storage and shipping facility that would be operated by the Army—and it was constructed very quickly in 1940. Wayne Enders, officer of the Parker County Historical Society, acknowledged that it displaced over 200 farms in a 22,000 acre area (W. Enders, personal interview), the western-most portion of which was about 5 miles from the city of Boomtown.

Much about the Arsenal had been on a grand scale: the construction project itself was considered the biggest of its kind to that point in history—the size of the campus, the number of buildings involved (over 1,000), the short time period in which it was constructed (in less than one year the area was cleared and at least one load line was constructed along with storage facilities)—and the production operations exceeded those of other installations (Arsenal repository documents). Indeed, it would produce more ammunition between 1942 and 1945 for the war effort than any of the other 59 installations in the country (Ohio History Central, 2008).

WAR-RELATED WORK, LABOR POOLS, AND TRAINING

Nelson (1995) notes that within the first few years of WWII, it was evident that the specialized production efficiencies that had been adopted by industry in the 1930s would need to change to accommodate a shortage of workers. Nelson observes that, "World War II production relied heavily on additional labor and a more elaborate division of labor . . . most assignments could be broken down to the point that prior industry or product-specific experience was unimportant" (p. 142).

Labor Market

Once the local labor market was exhausted within the first months of war, the Army went to all areas of the country to recruit workers for war-related work, and these people were bussed to the various sites (Atlas Powder Company, 1943). The Boomtown Arsenal employed between 15,000 and 19,000 people during WWII, and many of these came from West Virginia and the South. Ray McDaniel, a 42-year employee of the Arsenal, who was among those employed in its first years of operation, acknowledged that many recruiters actually targeted Appalachian females for war-related work, understanding that the men would then follow the women north (R. McDaniel, personal interview). Women did the light line work, while men did the heavier line work and pushed trays (R. McDaniel, personal interview). Prospective workers were bussed to the Arsenal for interviews and the signing of employment papers, and then moved on to training.

Appalachian migrants tended to be from West Virginia, while southern African Americans also were bussed into the area. Some found housing within the Arsenal's boundaries, but most located outside of the Arsenal. Appalachians tended to migrate to both Boomtown and Fieldview, while most African Americans tended to find housing in Boomtown.

Boomtown

As the closest city that had banks and stores where people could do their shopping, Boomtown drew many migrants and served as a social hub for workers. Troyer (1998) acknowledges that there were three USO clubs in Boomtown in addition to the bars that already existed there; and many bars would cash employment checks (p. 296). In fact, the city of Boomtown's population grew 15.5% in the 1940s, while that of Boomtown Township grew 25.7% by 1950 (p. 11). The Boomtown Planning Commission (1962) reported that "slightly over 52% of the adult area residents moved to the area since 1940 [and] almost 25% of total in-migrants came to Boomtown from other states (p. 9).

Fieldview

Most workers who stayed in Fieldview were Appalachians who rented in the government housing project in Apple Grove Park and from local residents. Very few African American migrants came to Fieldview, though those who did come were segregated, also housed in Apple Grove. The Apple Grove development was built by the government, and it is interesting to note how the government and Fieldview placed the literate institutions within Fieldview.

As the communities near the arsenals grew, the government invested in much of their infrastructure to accommodate the influx of workers and their families. The federal government provided funding to build or expand libraries in

Fieldview and neighboring Hartfield, and Fieldview's school district was also provided with funding to expand to accommodate the influx of workers. In fact, the federal and state governments have provided considerable funding to Fieldview since the Arsenal's original construction, developing much of the village's infrastructure and fire station. Furthermore, federal monies provided an infrastructure for Fieldview that could be used to facilitate literacy development in migrants and their children. These included construction of school buildings and a library.

Along Main Street, north to south, was the home/community of the workers; followed by the elementary schools; then the public library; then the high school and junior high school; then the railroad, which separates the homes, community, and school from the Arsenal, the workplace. Discussing institutional critique, Porter, Sullivan, Blythe, Grabill, and Miles (2000) describe the impact that institutions have in establishing spatial relationships within an organization. They write, "institutions . . . exercise power through the design of space (both material and discursive)" (p. 621). As the Army and Fieldview designed the community, they were able to consider the literacy infrastructure in addition to that of utilities and emergency services. They found that "literacy tends to be constructed in relation to the mandates of funding and policy interests (largely from government and industry) and to the goals articulated in large part by those interests" (p. 626).

The government was able to influence how the community was designed and where institutions were located because of the support—financial and economic—that the Arsenal brought. As the Army designed the community in which it would operate and house workers (Apple Grove), it included in the design an elementary school (Katie Thompson Elementary School). The elementary school is part of the community and it is physically located there. The high school building existed prior to the construction of the Arsenal, however the Army maintained the location and added to it to facilitate growth in population. The railroad tracks seem to represent a physical separation between the Fieldview community and the Arsenal, a transition to the workplace, which included its own literacy institutions—training programs and a library for employees.

Considering the physical relationship of how migrants were positioned in each community, Boomtown positioned migrants outside of the community while Fieldview, with more government support than Boomtown received, placed migrants within the community and offered material sponsorship for literacy.

TRAINING PROGRAMS AND LITERACY PRACTICES

Training can be discussed relative to assimilation of migrants into a new community and workplace training. This section first considers the training programs at arsenals around the country and then considers programs that each locale offered to migrants to assist in assimilation into the community. Because the government had considerable input into training and the literacy environment, it is also important to consider that input.

Government-Sponsored Training During WWII

With the migration of workers to areas near new arsenals, companies attempted to accommodate nonliteracy practices by relying more on oral and visual training programs. Also, to help migrants adjust to their new surroundings, the government, through the War Production Board and Office of Production Management, offered programs in the community. Generally, these took place at local schools or nearby colleges (Benson Polytechnic High School, 2006; UW-Extension, 2006). Rodabaugh (1975) indicates that "in Cleveland, programs were inaugurated during the war which eased African-Americans and Appalachian workers into the local labor market" (p. 313). These programs extended to Boomtown and Fieldview. St. John (personal interview) reported that church organizations had some programs to help Appalachians adjust to their new community; these included social gatherings and a community recreation program that sponsored sports leagues.

Also, Nelson (1995) states that, as the government recognized the problem of the mismatch between workers' skills and job requirements, "officials launched ambitious remedial efforts that became highlights of the mobilization experience [and] . . . the retraining of workers proved to be comparatively easy" (p. 143). As I noted before (Remley, 2009), historical documents from the Arsenal indicate that officials from the Arsenal attended a conference hosted by the Cleveland office of "Training Within Industry," which was a subdivision of the Office of Production Management. This conference helped generate a number of recommendations regarding the training of workers. Based on information about the training programs, these recommendations attempted to minimize the need for literacy skills among workers, understanding the transient nature of many of them (p. 99). This program is described more fully in Chapter 4 because it serves as the basis of the training experience for workers at the Arsenal and an illustration of a sponsorship of technical communication.

POPULATION PATTERNS IN AREA AFTER WAR

I close this chapter with an acknowledgment about population patterns after WWII in Fieldview to call attention to a phenomenon related to sponsorship of any type; what often happens when that sponsorship ends. I do so because it suggests a need for a smooth transition rather than an abandonment, as so often happens.

Fieldview's population has remained at the same level that it was during WWII, having not experienced much loss of the original influx of population into the area since WWII, nor growing much. Until 1960, the federal government contributed substantial levels of funding for development in the Fieldview area; however, the government sold the Apple Grove community to a private firm in 1954, resulting in the loss of that support, and the community deteriorated such that by the 1970s, Apple Grove was considered a poverty pocket within

Parker County (Lynott, 1989). The Arsenal closed its munitions operations after the Vietnam War, and several of the Arsenal's over 1,000 buildings have been burned down, while other areas within the Arsenal have been scaled down considerably. The Army currently maintains the area for National Guard training, and there is an effort to determine how to use the remaining space, much of which is contaminated by various toxic materials associated with building practices during the 1940s and with materials used to produce ammunition; this contamination negatively affects potential economic development in the area.

Because of the continued low-income housing infrastructure to the economy and the inability to expand industry in the area because of the Arsenal, Fieldview has the largest proportion of welfare recipients in Parker County (St. John, personal interview). This may reflect what Porter et al. describe as the ability of institutional structures both to "enable and discourage the progress of residents" (2000, p. 628). The Arsenal momentarily enabled progress by facilitating economic prosperity, but the temporary nature of its work (war-related) limited its ability to continue that progression, eventually turning into a space that discourages progress. This illustrates what can happen once sponsorship of programs ends, which is also evident today as cities experience weaker economies. This issue is discussed more in a later chapter. The next chapter, though, details the background of the Training Within Industry program mentioned earlier in this chapter.

APPENDIX C:
Great Arsenal of Democracy speech

Franklin D. Roosevelt Library and Museum Website; version date 2013. http://docs.fdrlibrary.marist.edu/122940.HTML

December 29, 1940

Radio Address of the President, Delivered from the White House

MY FRIENDS:

This is not a fireside chat on war. It is a talk on national security, because the nub of the whole purpose of your President is to keep you now, and your children later, and your grandchildren much later, out of a last-ditch war for the preservation of American independence and all of the things that American independence means to you and to me and to ours.

Tonight, in the presence of a world crisis, my mind goes back eight years to a night in the midst of a domestic crisis. It was a time when the wheels of American industry were grinding to a full stop, when the whole banking system of our country had ceased to function.

I well remember that while I sat in my study in the White House, preparing to talk with the people of the United States, I had before my eyes the picture of all those Americans with whom I was talking. I saw the workmen in the mills, the mines, the factories; the girl behind the counter; the small shopkeeper; the farmer doing his spring plowing; the widows and the old men wondering about their life's savings.

I tried to convey to the great mass of American people what the banking crisis meant to them in their daily lives.

Tonight, I want to do the same thing, with the same people, in this new crisis which faces America.

We met the issue of 1933 with courage and realism.

We face this new crisis—this new threat to the security of our nation—with the same courage and realism.

Never before since Jamestown and Plymouth Rock has our American civilization been in such danger as now.

For, on September 27th, 1940, this year, by an agreement signed in Berlin, three powerful nations, two in Europe and one in Asia, joined themselves together in the threat that if the United States of America interfered with or blocked the expansion program of these three nations—a program aimed at world control— they would unite in ultimate action against the United States.

The Nazi masters of Germany have made it clear that they intend not only to dominate all life and thought in their own country, but also to enslave the whole of Europe, and then to use the resources of Europe to dominate the rest of the world.

It was only three weeks ago their leader stated this: "There are two worlds that stand opposed to each other." And then in defiant reply to his opponents, he said this: "Others are correct when they say: With this world we cannot ever reconcile ourselves . . . I can beat any other power in the world." So said the leader of the Nazis.

In other words, the Axis not merely admits but the Axis proclaims that there can be no ultimate peace between their philosophy of government and our philosophy of government.

In view of the nature of this undeniable threat, it can be asserted, properly and categorically, that the United States has no right or reason to encourage talk of peace, until the day shall come when there is a clear intention on the part of the aggressor nations to abandon all thought of dominating or conquering the world.

At this moment, the forces of the states that are leagued against all peoples who live in freedom are being held away from our shores. The Germans and the Italians are being blocked on the other side of the Atlantic by the British, and by the Greeks, and by thousands of soldiers and sailors who were able to escape from subjugated countries. In Asia the Japanese are being engaged by the Chinese nation in another great defense.

In the Pacific Ocean is our fleet.

Some of our people like to believe that wars in Europe and in Asia are of no concern to us. But it is a matter of most vital concern to us that European and Asiatic war-makers should not gain control of the oceans which lead to this hemisphere.

One hundred and seventeen years ago the Monroe Doctrine was conceived by our Government as a measure of defense in the face of a threat against this hemisphere by an alliance in Continental Europe. Thereafter, we stood (on) guard in the Atlantic, with the British as neighbors. There was no treaty. There was no "unwritten agreement."

And yet, there was the feeling, proven correct by history, that we as neighbors could settle any disputes in peaceful fashion. And the fact is that during the whole of this time the Western Hemisphere has remained free from aggression from Europe or from Asia.

Does anyone seriously believe that we need to fear attack anywhere in the Americas while a free Britain remains our most powerful naval neighbor in the Atlantic? And does anyone seriously believe, on the other hand, that we could rest easy if the Axis powers were our neighbors there?

If Great Britain goes down, the Axis powers will control the continents of Europe, Asia, Africa, Australia, and the high seas and they will be in a position to bring enormous military and naval resources against this hemisphere. It is no exaggeration to say that all of us, in all the Americas, would be living at the point of a gun—a gun loaded with explosive bullets, economic as well as military.

We should enter upon a new and terrible era in which the whole world, our hemisphere included, would be run by threats of brute force. And to survive in such a world, we would have to convert ourselves permanently into a militaristic power on the basis of war economy.

Some of us like to believe that even if (Great) Britain falls, we are still safe, because of the broad expanse of the Atlantic and of the Pacific.

But the width of those (these) oceans is not what it was in the days of clipper ships. At one point between Africa and Brazil the distance is less from Washington than it is from Washington to Denver, Colorado—five hours for the latest type of bomber. And at the North end of the Pacific Ocean America and Asia almost touch each other.

Why, even today we have planes that (which) could fly from the British Isles to New England and back again without refueling. And remember that the range of a (the) modern bomber is ever being increased.

During the past week many people in all parts of the nation have told me what they wanted me to say tonight. Almost all of them expressed a courageous desire to hear the plain truth about the gravity of the situation. One telegram, however, expressed the attitude of the small minority who want to see no evil and hear no evil, even though they know in their hearts that evil exists. That telegram begged me not to tell again of the ease with which our American cities could be bombed by any hostile power which had gained bases in this Western

Hemisphere. The gist of that telegram was: "Please, Mr. President, don't frighten us by telling us the facts."

Frankly and definitely there is danger ahead—anger against which we must prepare. But we well know that we cannot escape danger (it), or the fear of danger, by crawling into bed and pulling the covers over our heads.

Some nations of Europe were bound by solemn non-intervention pacts with Germany. Other nations were assured by Germany that they need never fear invasion. Non-intervention pact or not, the fact remains that they were attacked, overrun, (and) thrown into (the) modern (form of) slavery at an hour's notice, or even without any notice at all. As an exiled leader of one of these nations said to me the other day, "The notice was a minus quantity. It was given to my Government two hours after German troops had poured into my country in a hundred places."

The fate of these nations tells us what it means to live at the point of a Nazi gun.

The Nazis have justified such actions by various pious frauds. One of these frauds is the claim that they are occupying a nation for the purpose of "restoring order." Another is that they are occupying or controlling a nation on the excuse that they are "protecting it" against the aggression of somebody else.

For example, Germany has said that she was occupying Belgium to save the Belgians from the British. Would she then hesitate to say to any South American country, "We are occupying you to protect you from aggression by the United States?"

Belgium today is being used as an invasion base against Britain, now fighting for its life. And any South American country, in Nazi hands, would always constitute a jumping-off place for German attack on any one of the other republics of this hemisphere.

Analyze for yourselves the future of two other places even nearer to Germany if the Nazis won. Could Ireland hold out? Would Irish freedom be permitted as an amazing pet exception in an unfree world? Or the Islands of the Azores which still fly the flag of Portugal after five centuries? You and I think of Hawaii as an outpost of defense in the Pacific. And yet, the Azores are closer to our shores in the Atlantic than Hawaii is on the other side.

There are those who say that the Axis powers would never have any desire to attack the Western Hemisphere. That (this) is the same dangerous form of wishful thinking which has destroyed the powers of resistance of so many con-quered peoples. The plain facts are that the Nazis have proclaimed, time and again, that all other races are their inferiors and therefore subject to their orders. And most important of all, the vast resources and wealth of this American Hemisphere constitute the most tempting loot in all of the round world.

Let us no longer blind ourselves to the undeniable fact that the evil forces which have crushed and undermined and corrupted so many others are already within our own gates. Your Government knows much about them and every day is ferreting them out.

Their secret emissaries are active in our own and in neighboring countries. They seek to stir up suspicion and dissension to cause internal strife. They try to turn capital against labor, and vice versa. They try to reawaken long slumbering racist and religious enmities which should have no place in this country. They are active in every group that promotes intolerance. They exploit for their own ends our own natural abhorrence of war. These trouble-breeders have but one purpose. It is to divide our people, to divide them into hostile groups and to destroy our unity and shatter our will to defend ourselves.

There are also American citizens, many of then in high places, who, unwittingly in most cases, are aiding and abetting the work of these agents. I do not charge these American citizens with being foreign agents. But I do charge them with doing exactly the kind of work that the dictators want done in the United States.

These people not only believe that we can save our own skins by shutting our eyes to the fate of other nations. Some of them go much further than that. They say that we can and should become the friends and even the partners of the Axis powers. Some of them even suggest that we should imitate the methods of the dictatorships. But Americans never can and never will do that.

The experience of the past two years has proven beyond doubt that no nation can appease the Nazis. No man can tame a tiger into a kitten by stroking it. There can be no appeasement with ruthlessness. There can be no reasoning with an incendiary bomb. We know now that a nation can have peace with the Nazis only at the price of total surrender.

Even the people of Italy have been forced to become accomplices of the Nazis, but at this moment they do not know how soon they will be embraced to death by their allies.

The American appeasers ignore the warning to be found in the fate of Austria, Czechoslovakia, Poland, Norway, Belgium, the Netherlands, Denmark, and France. They tell you that the Axis powers are going to win anyway; that all of this bloodshed in the world could be saved, that the United States might just as well throw its influence into the scale of a dictated peace, and get the best out of it that we can.

They call it a "negotiated peace." Nonsense! Is it a negotiated peace if a gang of outlaws surrounds your community and on threat of extermination makes you pay tribute to save your own skins?

Such a dictated peace would be no peace at all. It would be only another armistice, leading to the most gigantic armament race and the most devastating trade wars in all history. And in these contests the Americas would offer the only real resistance to the Axis powers.

With all their vaunted efficiency, with all their (and) parade of pious purpose in this war, there are still in their background the concentration camp and the servants of God in chains.

The history of recent years proves that the shootings and the chains and the concentration camps are not simply the transient tools but the very altars of modern dictatorships. They may talk of a "new order" in the world, but what they have in mind is only (but) a revival of the oldest and the worst tyranny. In that there is no liberty, no religion, no hope.

The proposed "new order" is the very opposite of a United States of Europe or a United States of Asia. It is not a government based upon the consent of the governed. It is not a union of ordinary, self-respecting men and women to protect themselves and their freedom and their dignity from oppression. It is an unholy alliance of power and pelf to dominate and to enslave the human race.

The British people and their allies today are conducting an active war against this unholy alliance. Our own future security is greatly dependent on the outcome of that fight. Our ability to "keep out of war" is going to be affected by that outcome.

Thinking in terms of today and tomorrow, I make the direct statement to the American people that there is far less chance of the United States getting into war if we do all we can now to support the nations defending themselves against attack by the Axis than if we acquiesce in their defeat, submit tamely to an Axis victory, and wait our turn to be the object of attack in another war later on.

If we are to be completely honest with ourselves, we must admit that there is risk in any course we may take. But I deeply believe that the great majority of our people agree that the course that I advocate involves the least risk now and the greatest hope for world peace in the future.

The people of Europe who are defending themselves do not ask us to do their fighting. They ask us for the implements of war, the planes, the tanks, the guns, the freighters which will enable them to fight for their liberty and for our security. Emphatically we must get these weapons to them, get them to them in sufficient volume and quickly enough, so that we and our children will be saved the agony and suffering of war which others have had to endure.

Let not the defeatists tell us that it is too late. It will never be earlier. Tomorrow will be later than today. Certain facts are self-evident.

In a military sense Great Britain and the British Empire are today the spearhead of resistance to world conquest. And they are putting up a fight which will live forever in the story of human gallantry.

There is no demand for sending an American Expeditionary Force outside our own borders. There is no intention by any member of your Government to send such a force. You can, therefore, nail—nail any talk about sending armies to Europe as deliberate untruth.

Our national policy is not directed toward war. Its sole purpose is to keep war away from our country and away from our people.

Democracy's fight against world conquest is being greatly aided, and must be more greatly aided, by the rearmament of the United States and by sending every ounce and every ton of munitions and supplies that we can possibly spare to help the defenders who are in the front lines. And it is no more unneutral for us to do that than it is for Sweden, Russia and other nations near Germany to send steel and ore and oil and other war materials into Germany every day in the week.

We are planning our own defense with the utmost urgency, and in its vast scale we must integrate the war needs of Britain and the other free nations which are resisting aggression.

This is not a matter of sentiment or of controversial personal opinion. It is a matter of realistic, practical military policy, based on the advice of our military experts who are in close touch with existing warfare. These military and naval experts and the members of the Congress and the Administration have a single-minded purpose—the defense of the United States.

This nation is making a great effort to produce everything that is necessary in this emergency—and with all possible speed. And this great effort requires great sacrifice.

I would ask no one to defend a democracy which in turn would not defend everyone in the nation against want and privation. The strength of this nation shall not be diluted by the failure of the Government to protect the economic well-being of its (all) citizens.

If our capacity to produce is limited by machines, it must ever be remembered that these machines are operated by the skill and the stamina of the workers. As the Government is determined to protect the rights of the workers, so the nation has a right to expect that the men who man the machines will discharge their full responsibilities to the urgent needs of defense.

The worker possesses the same human dignity and is entitled to the same security of position as the engineer or the manager or the owner. For the workers provide the human power that turns out the destroyers, and the (air)planes and the tanks.

The nation expects our defense industries to continue operation without interruption by strikes or lockouts. It expects and insists that management and workers will reconcile their differences by voluntary or legal means, to continue to produce the supplies that are so sorely needed.

And on the economic side of our great defense program, we are, as you know, bending every effort to maintain stability of prices and with that the stability of the cost of living.

Nine days ago I announced the setting up of a more effective organization to direct our gigantic efforts to increase the production of munitions. The appropriation of vast sums of money and a well coordinated executive direction of our defense efforts are not in themselves enough. Guns, planes, (and) ships and many other things have to be built in the factories and the arsenals of America. They have to be produced by workers and managers and engineers with the aid of machines which in turn have to be built by hundreds of thousands of workers throughout the land.

In this great work there has been splendid cooperation between the Government and industry and labor, and I am very thankful.

American industrial genius, unmatched throughout all the world in the solution of production problems, has been called upon to bring its resources and its talents into action. Manufacturers of watches, of farm implements, of linotypes, and cash registers, and automobiles, and sewing machines, and lawn mowers and locomotives are now making fuses, bomb packing crates, telescope mounts, shells, and pistols and tanks.

But all of our present efforts are not enough. We must have more ships, more guns, more planes—more of everything. And this can only be accomplished if we discard the notion of "business as usual." This job cannot be done merely by superimposing on the existing productive facilities the added requirements of the nation for defense.

Our defense efforts must not be blocked by those who fear the future consequences of surplus plant capacity. The possible consequences of failure of our defense efforts now are much more to be feared.

And after the present needs of our defense are past, a proper handling of the country's peacetime needs will require all of the new productive capacity—if not still more.

No pessimistic policy about the future of America shall delay the immediate expansion of those industries essential to defense. We need them.

I want to make it clear that it is the purpose of the nation to build now with all possible speed every machine, every arsenal, every (and) factory that we need to manufacture our defense material. We have the men—the skill—the wealth—and above all, the will.

I am confident that if and when production of consumer or luxury goods in certain industries requires the use of machines and raw materials that are essential for defense purposes, then such production must yield, and will gladly yield, to our primary and compelling purpose.

So I appeal to the owners of plants—to the managers—to the workers—to our own Government employees—to put every ounce of effort into producing these munitions swiftly and without stint. (And) With this appeal I give you the pledge that all of us who are officers of your Government will devote ourselves to the same whole-hearted extent to the great task that (which) lies ahead.

As planes and ships and guns and shells are produced, your Government, with its defense experts, can then determine how best to use them to defend this hemisphere. The decision as to how much shall be sent abroad and how much shall remain at home must be made on the basis of our overall military necessities.

We must be the great arsenal of democracy. For us this is an emergency as serious as war itself. We must apply ourselves to our task with the same resolution, the same sense of urgency, the same spirit of patriotism and sacrifice as we would show were we at war.

We have furnished the British great material support and we will furnish far more in the future.

There will be no "bottlenecks" in our determination to aid Great Britain. No dictator, no combination of dictators, will weaken that determination by threats of how they will construe that determination.

The British have received invaluable military support from the heroic Greek army and from the forces of all the governments in exile. Their strength is growing. It is the strength of men and women who value their freedom more highly than they value their lives.

I believe that the Axis powers are not going to win this war. I base that belief on the latest and best of information.

We have no excuse for defeatism. We have every good reason for hope—hope for peace, yes, and hope for the defense of our civilization and for the building of a better civilization in the future.

I have the profound conviction that the American people are now determined to put forth a mightier effort than they have ever yet made to increase our production of all the implements of defense, to meet the threat to our democratic faith.

As President of the United States I call for that national effort. I call for it in the name of this nation which we love and honor and which we are privileged and proud to serve. I call upon our people with absolute confidence that our common cause will greatly succeed.

CHAPTER 4

Training Within Industry: Sponsored Multimodal Technical Communication

In the previous chapter, I mentioned that representatives of the Arsenal attended a conference in Cleveland that featured discussion of the Training Within Industry program. The Training Within Industry (TWI) model was developed by the War Manpower Commission early in World War II to standardize training and some operations around the United States during WWII. As mentioned earlier, the philosophy developed by those who created the TWI program has been espoused in recent years as one that can lead to more efficient production methods, and it has been implemented in several companies, including Toyota, IBM, Micron Technologies, and New Balance, among others (Liker & Meier, 2007; TWI Learning Partnership, 2009). Understanding TWI in its original application is important to understanding it today, especially in terms of a discussion of the multimodal rhetoric associated with it and related sponsorship dynamics as a form of technical communication.

TWI OVERVIEW

TWI includes three principle programs, all of which begin with the word "Job," so they are called the "J-programs." These are Job Instruction, Job Methods, and Job Relations. Job Instruction details how to train workers, Job Methods focuses on improving performance and processes through innovation, and Job Relations addresses management of conflict between personnel. A fourth program is associated with the newer iteration of the program—Job Safety. This information originally was integrated into the others, however, as job safety programs developed, it became its own program within the TWI model. This chapter considers the multimodal rhetoric associated with all of these programs, but it emphasizes that rhetoric associated with the first two programs—Job Instruction and Job Methods—because these are the principle programs related to lean operation and continuous improvement. Modes of representation in these two programs can be studied readily based on documentation associated with them. Job Relations encompasses dynamics of management style and leadership skills, which emphasizes oral communication while integrating print-linguistic elements. Job Safety likewise involves oral/aural and print-linguistic skills similar to those described in this chapter.

Much of the information in this chapter is derived from the *Training Within Industry Report* (Dooley, 1945a) and some historical documents from the period 1941–1945, as well as the *TWI Workbook* (Graupp & Wrona, 2006) and *Training Within Industry: The Foundation of Lean* (Dinero, 2005). While these texts provide detailed discussion of applications of the programs, I focus discussion on modes used in training and process improvement. The TWI Report document serves as a primary source of information in understanding the specific programs that TWI implemented at plants throughout the country during WWII. Documents available at the TWI Service website are also reviewed, bridging some of the gaps between data points. The *Workbook* and Dinero's text help to understand how the TWI model is being applied today.

WWII and TWI

Even before U.S. involvement in WWII, the nation was manufacturing materials to support the Allies in their efforts to turn back German and Japanese imperialism. President Roosevelt provided some contextual background for the TWI program implicitly with his Great Arsenal of Democracy speech (December, 1940). I discussed the speech in the previous chapter, however, I call attention to its relationship to the TWI program here.

Specifically, Roosevelt explicitly acknowledged cooperation between government and industry in coordinating the massive effort toward war industry:

> Nine days ago I announced the setting up of a more effective organization to direct our gigantic efforts to increase the production of munitions. The appropriation of vast sums of money and a well-coordinated executive direction of our defense efforts are not in themselves enough. Guns, planes, ships and many other things have to be built in the factories and the arsenals of America. They have to be produced by workers and managers and engineers with the aid of machines which in turn have to be built by hundreds of thousands of workers throughout the land. In this great work there has been splendid cooperation between the government and industry and labor. And I am very thankful.

In this portion of the speech, he articulated the effort to minimize resistance that could have occurred in hiring people to work in the munitions plants in an effort to maximize potential production and the coordination of government and industry to sponsor that effort.

Many skilled tradesmen enlisted in the military, and many people who remained were more familiar with farming skills than with industrial skills. The

War Manpower Commission, in conjunction with industry leaders, developed the TWI program, and representatives held conferences around the country to help munitions plants implement it. TWI attempted to minimize the need for literacy skills among supervisors and workers, understanding their varied literate backgrounds and their transient nature.

TWI's founding principle was that "'What to do is not enough.' It is only when people are drilled in 'how to do it' that action results" (Dooley, 1945a, p. xi). The most basic objective of the program was to train supervisors and workers quickly to produce defense materials in order that such training, efficient in itself, would result in efficient and quality production. The report acknowledges that

> the training we give the worker . . . can be more than an expedient means of getting the job done. It can be suitable to the individual and in line with his native talent and aspiration. Then it becomes education because the worker . . . trained in accordance with his talent and aspiration, is a growing individual. (Dooley, 1945a, p. xii)

An important consideration in Mayer's (2001) theory is the accounting for individual differences based on learners' experiences and knowledge prior to receiving instruction. Kress and van Leeuwen (2001) posit that production of meaning includes "experiential meaning potential." They state that, "humans have the ability to match concepts with appropriate material signifiers on the basis of their physical experience of the relevant materials" (p. 75). Learning tends to occur by doing, and the more experiences one has with a given set of procedures, the more he or she can understand certain routines. Such experiences help develop a mental map of the system associated with the process, the TWI program emphasizes these attributes of semiotics.

JOB INSTRUCTION

Within TWI, Job Instruction covers processes to help train frontline workers to perform a given task and to train supervisors to perform their tasks. TWI developed this program to quickly train workers by breaking down jobs into simple processes (Dooley, 1945a, p. 21). This is consistent, actually, with the task allocation philosophy of Taylorism. In the late portion of the 19th century and early part of the 20th century, Frederick Taylor implemented an approach to management that became known as Taylorism. One principle of Taylorism that has raised considerable debate is that of "task allocation," in which experts, who comprise management, break down a process into smaller processes such that unskilled workers can manufacture the product in stages (Taylor, 1913, p. 38). Such allocation reduces the need for highly trained workers to be hired while improving efficiency. Given the demand for unskilled labor at the time, as I indicated in the previous chapter, this philosophy was part of the TWI method.

Further, given that many who came to work in the arsenals around the country were illiterate or of low literacy levels, as indicated in the previous chapter, TWI encouraged an emphasis on visual modes for training purposes. After operations were broken down into simpler tasks, training on each task would include the following instruction:

1. Show him how to do it
2. Explain key points
3. Let him watch you do it again
4. Let him do the simplest parts of the job
5. Help him do the whole job
6. Let him do the whole job—but watch him
7. Put him on his own (Taylor, 1913, pp. 19, 21).

TWI also developed a package of training programs that developed supervisory instruction as well as job instruction programs (Taylor, 1913, p. 30).

Generally, the following four essential elements were presented to plants:

1. The training program should be one of utter simplicity.
2. It must be prepared for presentation by intensive and carefully "blue-printed" procedure, utilizing a minimum of time.
3. It must be built on the principle of demonstration and practice of "learning by doing," rather than on theory.
4. The program should provide for "multipliers" to spread the training by coaching selected men as trainers who, after being qualified in an institute . . . pass the program on to supervisors and their assistants who would use it in training men and women workers. (Taylor, 1913, p. 32)

The principles above express the design, production, and distribution dynamics of the training that is encouraged: Item number 3 articulates design (demonstration and practice), and item number 4 offers suggested production and distribution dynamics—multiplier effect by trained supervisors. This training emphasizes the visual, aural, and gestural/behavioral (through practice). These attributes are also included in multiple levels of training. I described the program's original implementation, which included using films, narration, models, supervisory training "which included instruction on how to teach a job, management principles in regard to human relations, and later, instruction on how to improve the method of doing the job" (Remley, 2009, p. 104). However, I provide much more detail about the program here.

As Mayer (2001) asserts, presenting text and image simultaneously enables the learner "to hold mental representations of both in working memory" (p. 96). Baddeley (1986) acknowledged that there is a phonological (auditory) channel and a "visuo-spatial" (visual) channel associated with short-term memory. Schnotz (2005) suggests that when a visual image is presented to a reader, the reader can create a visual model as he/she listens to a narrative about the picture. The reader is forced to process the words while also trying to develop a mental model of the concept or activity if only text is provided. This creates an overload in working memory and compromises the ability to learn (pp. 54–55). People can better process information if both channels are used than they can when too much of one system is used.

Part of the visual, aural, and spatial representations in demonstrations is the trainer's behaviors and voice. Supervisors were encouraged to use the original instructional materials as an outline, but eventually a script was developed that included stage directions, including the behaviors shown in Figure 4.1.

The trainers were taught how to perform their demonstrations in such a way that would create a certain effect on the trainee. This may help the trainee remember certain emphases or the importance of certain steps. As Dinero (2005) notes,

> A trainer must know the manual's content like a stage director knows the dialogue, cues, action and intent of a script, and then be able to deliver the lines like an actor. Like a good actor, a trainer makes participants believe what he or she is saying and doing. (p. 156)

JOBS in YOUR departments."

— Explain you are "taking apart" the process of instruction and examining each part separately.

— Select a member near you: Do not ask him to stand. Turn to him. Address him personally.

— Ask:

— "Do you know how to tie the knot?" (If he knows, turn to another.)

— "Let me TELL you how to tie a fire underwriters' knot. Listen closely."

Note to Trainer: Put your HANDS IN YOUR POCKETS: Have wire in table drawer or your pocket. Don't have wire in sight.

Figure 4.1. TWI training script.
Document courtesy of the U.S. Army.

Mental mapping is among the attributes articulated by Senge (2006) as important in today's workers. Pinker (1997) explains that mental representations can occur in various modes: visual, auditory, print-linguistic, and "mentalese" (pp. 89–90). How one comes to their own mental map of a given concept depends on how that person has experienced a given phenomenon previously and is able to generate a mental abstraction of a concept. Mental maps help a person to understand new experiences through the engagement of prior experiences related to the new experience and relative to modes with which they have interacted previously relative to those experiences.

The documents associated with TWI's instruction and improvement models illustrate features of mapping using combinations of multiple representation systems.

Manuals

Manuals were used primarily as reference materials, though once a worker completed training, they were very familiar with their job. One may assume that the entire operation may not be effectively represented textually or through pictures or a combination. Amerine and Bilmes (1990) state that instructions may assume a certain embodied knowledge possessed by the user. While information is not explicitly represented textually or graphically, certain attributes of the operation may be implied (p. 327). The opportunities to practice tasks after demonstrations offered employees embodied knowledge.

Supervisor Training

Visual modes of representation were emphasized not only in training the lower level workers, but also supervisors. Because demand for munitions workers exhausted the existing number of trained workers, the government recognized that they would need to train new supervisors. TWI had a program in place for such training that revolved around addressing what it identified as five particular needs of supervisors. These are "Knowledge of the Work, Knowledge of Responsibility, Skill in Instructing, Skill in Improving Methods, and Skill in Leading" (Dooley, 1945a, p. 48).

As part of their orientation, supervisors viewed storage charts and then toured the entire facility (Atlas Powder Company, 1943, pp. 273–274). Such an orientation provides a visual understanding of how the different operations are situated in the facility. This tour offers visual as well as gestural/behavioral modes of representation. Through the experience of moving from one location to another, one becomes familiar with spatial relationships they will encounter in the actual practice. Another feature was the use of miniature models in the training for storage, loading, and blocking of ammunition (Atlas Powder Company, 1943, p. 270). This represents another example of visual and behavioral modes to facilitate an understanding of spatial relationships associated with the processes.

Of fieldwork, the SOO from 1942 explains that "it was felt that this was a minimum in which they could absorb, through actual handling, sufficient knowledge of the methods and procedures involved in handling the enlarged variety of ammunition now is use" (Atlas Powder Company, 1942, p. 267). Again, this is an example of visual and behavioral modes of representation. The same document acknowledges, further, that

> demonstrations as to the proper method of storing ammunition in maga-zines and of loading and blocking it in railroad cars were given by the instructors. In order to carry out these demonstrations miniature models of igloos, railroad cars, and several types of ammunition were designed and constructed to scale, thus making it possible to follow the specifications as given on the loading and storage charts. After the demonstrations had been completed, students were given ample opportunity to inspect and, later, to practice with the miniatures. (p. 270)

Frontline Worker Training

Supervisors did the primary initial training of workers hired into the facility. Figure 4.2 shows how the process they used is described in historical documents.

Clearly, the various modes worked to reinforce the information conveyed in each. Even the actual demonstrations integrated real models and materials in addition to the oral and visual attributes of demonstrating the task. Dinero (2005) explains that the trainer talked through the steps as he showed trainees the task. Then, workers were given a chance to do the task themselves with the trainer's supervision. Such training engages all of the modes listed by the New London Group (1996).

JOB METHODS

As workers and supervisors perform their jobs, they are encouraged to think about any weaknesses in the process that could be improved upon. I mentioned above that to help employees with limited industrial skills learn their job quickly, the TWI program broke jobs down such that a job that could be done by a single skilled tradesman, who had some years of experience and training, could be performed, instead, by three or four workers who learned their tasks quickly. While appearing to be inefficient because of the additional workers used, it took very little time to train each to perform their respective job. In breaking the job down into multiple parts, TWI encourages supervisors to itemize specific steps. This facilitates analysis of the system and potential improvements to it.

Supervisors are encouraged to break the job down into its different parts, analyze each step, and consider if any step is wasteful. For this activity, they write each of the steps in print-linguistic form, generating a narrative of the operation. They also develop a diagram that represents the system visually to help visualize

The Fuze and Booster production lines showed a need for vestibule training of operators as these were fairly complicated assembly operations, some of which involved loaded component parts. It was especially necessary during the period in early 1942 when 25 to 50 new employes were being added daily.

At first, movies showing the loading and assembly of M-48 fuzes, anti-aircraft shells, etc. were shown after working hours to the inter-viewers and other personnel connected with employment. The M-48 fuze film was then shown to new employes going to work on the fuze lines. The pro-cedure was to give the Induction and Safety Talks to all new employes at 8:30 a.m. and then take them, with the exception of the fuze line employes, to another area to be fitted for uniforms and safety shoes. The fuze line employes remained in the Conference Room for the movie on "Loading and Assem-bly of the M-48 Fuze." When the movie was finished, these employes had their lunch after which they also were taken out to be fitted for uniforms and powder shoes. As the M-48 picture was a silent film, a narrative was pre-pared explaining the various scenes.

Figure 4.2. Description of training.
Courtesy of the U.S. Army.

the entire process. As they generate a new process to replace the existing one, they do the same things—break the job down and represent it textually with both print-linguistic and graphic representations. Then, to propose a new process, they must develop a document that integrates print-linguistic text as well as visuals.

Job Methods: Historical Application

Figures 4.3 and 4.4 show breakdown charts used in the TWI program. The examples are taken from a 1944 set of documents, however, the charts are very similar today (see Graupp & Wrona, 2006). Figure 4.3 shows the "current" process, while Figure 4.4 shows the "proposed" method.

JOB METHODS BREAKDOWN SHEET

Operation <u>Inspect, Assemble, Rivet, Stamp and Pack</u> Product <u>Radio Shields</u> Department <u>Riveting and Packing</u>

Your name <u>Bill Brown</u> Operator's name <u>Jim Jones</u> Date <u>June 14, 1944</u>

List of All Details for { PRESENT / PROPOSED } Method Every single thing that is done—Every inspection—Every delay	NOTES Reminders—Tolerances—Distance—Time Used—Etc.	IDEAS Write them down—Don't trust your memory
1. Walk to box of copper sheets.	Placed 6 feet from bench by handler.	
2. Pick up 15 to 20 copper sheets.		
3. Walk to bench.		
4. Inspect and lay out 12 sheets.	Scratches and dents. Scrap in bins.	
5. Walk to box and replace extra sheets.		
6. Walk to box of brass sheets.	Placed 3 feet from copper box by handler.	
7. Pick up 15 to 20 brass sheets.		
8. Walk to bench.		
9. Inspect and lay out 12 brass sheets.	One on top of each copper sheet.	
10. Walk to box and replace extra sheets.		
11. Walk to bench.		
12. Stack 12 sets near riveter.		
13. Pick up one set with right hand.		
14. Line up sheets and position in riveter.	Line-up tolerance .005".	
15. Rivet top left corner.		

(OVER)

W-810-1

Figure 4.3. Current process.
Document courtesy of the U.S. Army.

JOB METHODS BREAKDOWN SHEET

Operation___Inspect, Assemble, Rivet, Pack_____   Product__Radio Shields_____   Department__Riveting and Packing___

Your name___Bill Brown_____   Operator's name_____   Date__June 14, 1944___

List of All Details For (Present / Proposed) Method	NOTES	IDEAS
Every single thing that is done—Every inspection—Every delay	Reminders—Tolerances—Distance—Time used—Etc.	Write them down—Don't trust your memory
1. Put pile of copper sheets in right jig	Boxes placed on table by Handler	
2. Put pile of brass sheets in left jig		
3. Pick up 1 copper sheet in right hand and 1 brass sheet in left hand		
4. Inspect both sheets	Scratches and dents. Drop scrap through slots	
5. Assemble sheets and place in fixture	Fixture lines up sheets and locates rivet holes. Brass sheet on top	
6. Rivet the 2 bottom corners		
7. Remove, reverse, and place sheets		
8. Rivet the 2 top corners		
9. Place Shield in front of fixture Repeat No. 3 to No. 9 incl.—19 times		
10. Put 20 Shields in shipping case— 200/case	Cases placed by Handler	
11. Carry full cases to Packing Dept.	By Handler with hand truck	
12. Close, weigh, and stencil cases	Check inspection by Packer	
13. Write weight on delivery slip		
14. Set cases aside for shipment		

16—37207-1   U. S. GOVERNMENT PRINTING OFFICE

Figure 4.4. Proposed process.
Document courtesy of the U.S. Army.

While formatted as a list rather than as a paragraph, the form helps identify steps that may cause delays or problems. The notes about each step act as a means of evaluation embedded within the accounting of the steps, thereby integrating the reflective attributes of narrative in addition to the detailed sequencing of specific steps. A diagram of the steps provides a visual representation of the same "current" process (Figure 4.5).

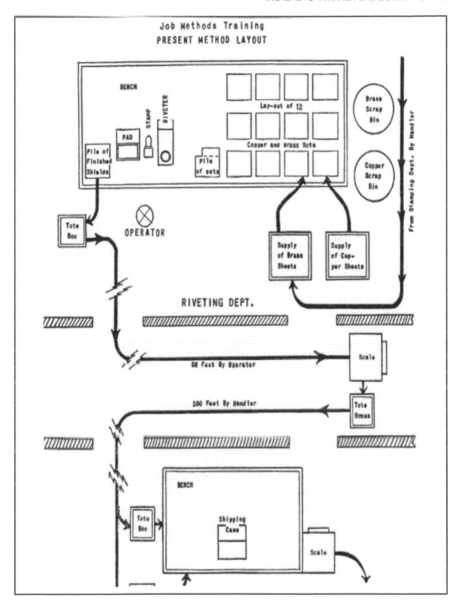

Figure 4.5. Diagram of current process.
Document courtesy of the U.S. Army.

The breakdown sheet offers a narrative of the steps associated with the task, but the diagram shows what it looks like as a system. These are repeated in the documents associated with the proposed process.

Including both a table with print-linguistic text to list the steps and notes or reasons related to them as well as drawing a diagram of the process in each representation (existing and proposed) combines benefits of multimodal representation to help the reader understand not only what is occurring but what that looks like systematically. The diagram is able to show the system in process.

Finally, the proposal, which may include the breakdown sheets and diagrams as supplemental materials, would be written and submitted, integrating a summary as well as a budget/cost comparison for the processes. The persuasive elements of the proposal include the acknowledgement that "material scrap" will be reduced "from 15% to less than 2%" and the new process will use fewer "inexperienced operators" than the existing process does. While these examples are taken from the 1940s, today's forms are similar, integrating additional categories to further code the data/steps to facilitate analysis. For example, one additional category is "Remarks," which is somewhat similar to the "Notes" category in the 1944 sheet. With that category one may code a given step relative to the time it takes to perform it, how it makes the operation safer or efficient or other attributes of the particular step that contribute to considerations of whether to eliminate or keep it (see Dinero, 2005, p. 260 and Graupp & Wrona, 2006, pp. 89–115 for examples).

THE TWI BALANCING ACT

The New London Group states that, "in some cultural contexts . . . the visual mode of representation may be much more powerful and closely related to language than 'mere literacy' would ever be able to allow" (1996, p. 64). However, the Group also observes the dynamic nature of language and representation. TWI enables workers to form mental maps of concepts and think systematically about concepts and processes. It encourages use of visual and print-linguistic text toward improved cognition of concepts and processes. The Job Methods sheets, also, facilitate analysis using a visual form of representation that includes print-linguistic text.

These modes of representation act to enhance and reinforce each other cognitively when used in various combinations. According to Arnheim (1969), when an image or diagram is provided, "the mind visualizes the whole image, whereas text must piece the image together through a linear process" (p. 249). The diagram helps develop a mental map of the system, which contributes to visualizing and analyzing potential changes to the system and how those may affect other parts of the system.

The TWI program helped to train lineworkers quickly by emphasizing visual modes of representation, and it developed supervisors by giving them a variety of skills, including the ability to train workers, as described above. Employees in the TWI program are exposed to multiple modes of representation—visual, behavioral, auditory, oral, and print-linguistic. Different combinations of these modes are evident across different levels in the organizational hierarchy, yet they all contribute to helping readers or viewers understand each process as a system and draw on mental mapping and the cognitive abilities thereof to improve processes.

CONCLUSION

This chapter has provided insight into the origins of TWI and its applications. However, much of this information is based on the TWI Report (Dooley, 1945a), which was published *after* the war. The implementation of the TWI program had experienced its growth and worked out some flaws by that time. The next chapter describes an implementation of it prior to the publication of the Report, showing the evolution of TWI's materials and rhetoric relative to a particular case study. In particular, a catastrophic event contributed to changing some elements of how it was applied, and these changes include modifying some elements of multimodal rhetoric that it includes. These changes contribute to the sponsorship dynamics associated with TWI.

CHAPTER 5

Training Practices, the Accident, and Sponsorship Implications

That's one thing Mr. Smith, who was my boss, was very adamant about. Would walk around and lift his arms, "if there's anyone who cannot spell "personnel," they best not be in this office."

—Interview participant

Different literacy skills were required to work at different levels within the Arsenal's organization. The quotation above foreshadows some of the findings and implications related not only to the accident and evolution of materials but also to literacy sponsorship. One's literacy skills could affect what work employees at the Arsenal could do in spite of certain accommodations for low literacy levels. It was also clear that certain literacy skills were esteemed while others were withheld. Most importantly, there is evidence that a failure at the workplace to balance effectively those literacy practices that were esteemed with those associated with the accommodations made for employees with low literacy levels contributed to the bomb accident that resulted in the death of 11 employees as well as destruction of a storage igloo and damage to several other buildings.

As I have acknowledged previously, Brandt (2001) defines sponsors of literacy as "any agents, local or distant, concrete or abstract, who enable, support, teach, and model, as well as recruit, regulate, suppress, or withhold, literacy—and gain advantage by it in some way" (p. 19). I specify those agents and their relationship to each other and the community. Then I present information about training practices—training that predates the accident, training just after the accident, and training several years subsequent to the accident. The practices prior to the accident contributed to the accident, and the revisions made to training materials and practices subsequent to the accident addressed those issues.

THE MAJOR SPONSORS OF LITERACY

Three agents act as principle sponsors during the period of study: the government, Atlas Powder Company (APCO), and Vulcan Tires. As were all of the arsenals built for WWII, the site is operated as a "Government-Owned,

Contractor-Operated" (GOCO) site. APCO was the contractor during WWII, and shortly after the war ended in 1945, the government, through the War Department, took it over entirely to store munitions not used in the war. Vulcan Tires was hired by the government in 1952 as the operating contractor as U.S. involvement in the Korean War increased, and Vulcan Tires took over operations from the government shortly after the war ended. The government still owned the site, but the War Department no longer operated it.

Because the Arsenal was established through a government-funded effort to create arsenals throughout the country to produce weapons for the Allies, the government was the primary sponsor during WWII and until Vulcan Tires assumed control in 1952. APCO was a co-sponsor during WWII, but the government, especially the Army, was very much involved in operations and seems to have had a greater role in decision-making processes.

Generally, a variety of literacy practices were used at the Arsenal, covering the range identified by the New London Group (1996): print-linguistic texts, aural, visual, spatial, gestural, and multimodal. Also, different forms of documents required different kinds of literacy skills, and there appears to have been a literacy-related hierarchy in which more print-linguistic skills were required of employees at higher levels. The investigation report related to the accident offers insight into these practices as well as implications related to the separation of print-linguistic practices and oral/visual/experiential practices.

THE ACCIDENT

Description of Event

On March 24, 1943, an explosion in the depot area of the Arsenal killed 11 people and injured another 3. The explosion destroyed an entire storage igloo as workers were unloading and moving boxes of small, 20-pound fragmentation bombs (see Figure 5.1). The investigation concluded that the combination of a defect in the fuse used to detonate the bombs and the rough handling by workers, in spite of expressed precautions that pertained to the defect, caused the explosion to occur (Stratton, 1943, p. 13).

Report of Investigation Into Accident

In the investigation report, Stratton (1943) acknowledges various attributes of practices that depot workers used when unloading and moving bombs into position within storage igloos. Employees received specific training but moved away from some attributes of that training in actual practice.

Based on interviews with workers, among a list of findings, Stratton (1943) acknowledges the routine procedure for unloading the bombs. He acknowledges a change in procedure as more boxes were unloaded or as workers perceived the need to unload quickly:

BOMB, FRAGMENTATION, 20-LBS, AN-M41 & AN-M41A1

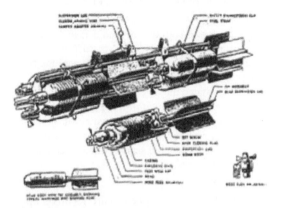

Body. This bomb is constructed of cast-steel nose and tailpieces, a seamless steel inner tube, and a helically wrapped drawn steel wire wrapping around the inner tube. The tube is threaded to hold the nose and tail section

Suspension. For individual suspension of this bomb, a U-shaped eyebolt of steel is welded to the body at the center of gravity for horizontal suspension, and an eyebolt is welded to the tail for vertical suspension. The bomb may be dropped in a cluster of six bombs in the *Cluster Adapter AN-M1A2 or M1*, forming the *Cluster AN-M1A1 or M1.* The cluster adapter is made of sheet steel, and does not use eyebolts of bombs for suspension.

Tail. Four rectangular sheet-steel vanes are welded to a length of one-inch cast-iron pipe which screws into the base-filling plug.

Over-all length	19.5 inches
Body length	11.3 inches
Diameter	3.6 inches
Over all weight	20.3 pounds
Filler	TNT
Filler weight	2.7 pounds
Fuzing	M158, AN-M110A1, M110, M109

Figure 5.1. Diagram and specifications of fragmentation bomb. Document courtesy of the U.S. Army.

f. The routine procedure noted above follows. The railroad car is opened. The door bracing and bulkhead bracing are removed. The semi trailer is backed to the door- opening, and beginning with the outside box of the upper tier the boxes are handed down and placed in the truck. This first unloading operation requires the boxes to be handled by two men but as more boxes are removed from the car it was common practice for one man to pull a box from the tier: (the box and contents weigh 168 pounds), permit it to slide down through his arms on to the floor and then to walk the box on its corners to the truck, where it would be taken by two stackers and placed in position. This procedure was continued until a trailer load was completed. (p. 7)

Workers were trained to unload the boxes as a two-man operation. However, the workers modified it so that only one worker would do the job of two, resulting in dangerous handling of the boxes. Stratton acknowledges that

> the handling of the boxes at both the railroad cars and the igloo was carried out according to usual procedure. Helper No. 3070 working in the railroad car substantiated the previous testimony but indicated the boxes containing the bomb clusters were difficult to handle inasmuch as they were not provided with handles. He stated they were instructed to give special care to this shipment as it was the first shipment of this type of munition that had been received at the Depot. He detailed the procedure of unloading the cars as described in paragraph 5 above, making the statement that the boxes were permitted to slide down between the arms of the workers, landing on the end, and then were walked to the truck. (1943, p. 11)

In spite of being "instructed to give special care to this shipment," the workers used a procedure other than the one associated with their training, which compromised safety.

Much of the training at the Arsenal came in the form of visual, aural, and experiential training. Workers were shown how to do a given task through demonstrations and then they practiced doing the task. In some cases, very little time passed between the start of training and the worker assuming work on the task. This is discussed further later.

Within a listing of interviews, Stratton also acknowledges that the fuse assembly of the bombs involved was new and defective in its tolerances:

> f. During the discussion, Captain Dorsey of the Kingsbury Ordnance Plant showed data which indicated fuze M-110 had had an extremely bad loading history . . . over one million of these fuzes have had to be re-worked before they could be made sufficiently reliable to be placed on the bombs. The chief weakness in the fuze is in the pinion column which when coming out of adjustment generally permitted the safety blocks to fall out and the fuze to become armed. This data report in the form of a memorandum to O. E. Ralston dated January 6, 1943, indicating the defects in these fuzes, will be requested from Kingsbury Ordnance Plant. (1943, p. 11)

This passage indicates not only that a defect that could endanger the lives of others was known prior to the accident, but that it was conveyed to others via a memo—a print-linguistic document.

While workers received their training mostly through visual, aural, and experiential practices, administrators often communicated through print-linguistic forms of literacy. Print-linguistic materials were available to workers, however, how they used those materials is another matter. Also, literacy expectations across these materials and positions varied considerably. Training practices affect how employees perform certain tasks, and these practices include the various ways they are taught about performing those tasks. There are clear differences in literacy practices relative to one's position and relative to the accident's timing.

The information in the investigation report shows that workers were not told adequately how to be careful handling the boxes despite officers' knowledge about the need for such care. A memo that specifies problems with the fuse and related hazards exists before the accident, and testimony shows that workers were told to be careful. However, their use of the dangerous one-man protocol suggests that they did not know exactly why they needed to be careful or how to be careful. A breakdown in communication occurred somewhere between the original memo and the instruction to the workers to be careful. This breakdown suggests that certain information ought to be conveyed in multiple modes of representation—orally and in writing—to reinforce each other. Indeed, Stratton concludes the report with the acknowledgement that

> b. Upon verbal report of the undersigned to his commanding officer, action was initiated to locate all other shipments of twenty pound fragmentation bombs M-41 in transit or storage and to caution all authorities that extreme care must be taken in the handling of these bomb clusters in shipping boxes. (1943, p. 14)

This cautioning of "all authorities" evidently includes supervisors; because the Ammunition, General (1945), used by supervisors in their training, includes explicit references to precautions associated with rough handling or dropping of boxes, which are highlighted with italics to call attention to them (Figure 5.2).

---

g. Fuzes.

(1) *Extreme care must be taken in handling and assembling fuzes to shell or bombs. All fuzes must be treated as delicate mechanisms. .The forces which arm a fuze on firing a weapon can be simulated by rolling or dropping, and a fuze so armed may be functioned by the impact of a blow or by dropping.*

Figure 5.2. Precautions in manual published after accident (p. 303).
Courtesy of the U.S. Army.

No such references occur in manuals that predate the accident. This suggests that such references were part of the changes to the print materials recommended as a result of the accident. A relationship between visual, aural, experiential, and print-linguistic literacies exists, and revisions to documents reflect this relationship.

LITERACY PRACTICES REQUIRED IN THE WORKPLACE'S TRAINING PROGRAM

Virtually all employees of the Arsenal experienced some part of the training program, no matter the level at which they worked; consequently, it is important to consider various attributes of the literacy sponsorship associated with that training. The primary training workers received came in the form of hands-on training, wherein lineworkers learned how to do their job by actually watching trainers perform the task and then doing it themselves under some supervision. This was part of the Training Within Industry (TWI) program described in the previous chapter.

Pre-Accident Training—Lineworkers

"Roger," who began work at the arsenal the earliest (1942), reported that training was heavily visual and hands-on in this interview exchange:

> Me: Do you remember the training that you received when you begin, first began work at the arsenal?
>
> R: Very little.
>
> Me: Very little. Do you remember? Very little recollection of it or very little training?
>
> R: You pick up a shell, put it in a vice
>
> Me: And they showed this to you, or you didn't have anything to read but they showed you how to do this?
>
> R: No, no. It was very . . .
>
> Me: Hands-on?
>
> R: Hands-on. Very hands-on

Training of lineworkers emphasized visual, oral, and experiential literacies. I asked Roger specifically if there had been any reading of print materials in this training, and he acknowledged that there had not. Archived documents confirm this reliance on visual and hands-on training.

According to its 1942–1943 Summary of Operations, the training programs offered at the Arsenal used multiple modes of presentation. Principally, the training programs relied on visual modes, although manuals were distributed to employees. Mayer (2001) observes that multimodal presentation is most effective when low-knowledge learners (those with little previous experience with or knowledge about the task) and/or high-spatial learners (those who have the ability to process spatial information quickly, also visual learners) are involved (p. 161). The Arsenal's History of Operations (Atlas Powder Company, 1943) acknowledges that, "training was concentrated upon two principal centers of activity: 1) employe training, which included an induction talk to all employes and pre-employment or vestibule training, lasting two or three days, for fuze and detonator line operators" (p. 110).

The induction talk was given by the employment interviewer. "Induction talks were given to over 19,000 employes from December, 1941 until April 1942." (Atlas Powder Company, 1943, p. 110). As seen in the excerpted description below (taken from page 110), vestibule training included the showing of silent movies that were accompanied by narration (Figure 5.3).

At first, movies showing the loading and assembly of M-48 fuzes, anti-aircraft shells, etc. were shown after working hours to the interviewers and other personnel connected with employment. The M-48 fuze film was then shown to new employes going to work on the fuze lines. The procedure was to give the Induction and Safety Talks to all new employes at 8:30 a.m. and then take them, with the exception of the fuze line employes, to another area to be fitted for uniforms and safety shoes. The fuze line employes remained in the Conference Room for the movie on "Loading and Assembly of the M-48 Fuze." When the movie was finished, these employes had their lunch after which they also were taken out to be fitted for uniforms and powder shoes. As the M-48 picture was a silent film, a narrative was prepared explaining the various scenes.

Figure 5.3. Description of induction.
Document courtesy of the U.S. Army.

Roger's comment and this passage indicate that training of lineworkers emphasized visual, aural, and experiential literacies. Training for those who worked in the depot area also emphasized visual, aural, and experiential literacies. The same document acknowledges that

> demonstrations as to the proper method of storing ammunition in magazines and of loading and blocking it in railroad cars were given by the instructors. In order to carry out these demonstrations miniature models of igloos, railroad cars, and several types of ammunition were designed and constructed to scale, thus making it possible to follow the specifications as given on the loading and storage charts. After the demonstrations had been completed, students were given ample opportunity to inspect and, later, to practice with the miniatures. (Atlas Powder Company, 1944b, p. 270)

As I described in the previous chapter, this form of training was developed by the government in conjunction with industry. Recognizing that many workers had little industrial training, particularly related to the war industry, the government and industry leaders developed the Training Within Industry (TWI) program. This program's learning model was based on the adage that one learns through experience (Dooley, 1945b). According to the *TWI Report* (Dooley, 1945b), "People have to learn to do jobs . . . Learning by doing is good, planned training" (p. 17).

Visual, aural, and experiential modes of presentation are evident in the description of the instruction. Mayer (2001) acknowledges that it is important to present both pictures and words simultaneously rather than in succession. Presenting them simultaneously enables the learner "to hold mental representations of both in working memory" (p. 96). Moreover, the worker is learning how to do a specific task in the context that he or she will be doing it. While not mentioned explicitly in the *TWI Report*, this is a form of experiential education, which encourages learners to apply skills they are trying to learn through situated practice under a teacher's supervision (Dewey, 1938).

Pre-Accident Training—Supervisors

While lineworkers' training emphasized visual and hands-on forms of practices, supervisor training integrated more print-linguistic materials in addition to visual and hands-on practices. Because of the depleted labor market, the government developed the TWI program. As mentioned in the previous chapter, TWI identified five particular needs of supervisors: knowledge of work; knowledge of responsibilities; skill in instructing; skill in improving methods; and skill in leading (War Manpower Commission, 1945, p. 48).

While lineworker training lasted only a few days, supervisor training lasted 8 weeks, considering the time needed to train in the five needs areas identified above. Also, while most of the instruction portion included explanation of visual diagrams and demonstrations of production procedures, half of it (144 hours) involved fieldwork. Relative to the time spent in fieldwork, the Historical Report of Operations from (Atlas Powder Company, 1942) explains that "it was felt that this was a minimum in which they could absorb, through actual handling, sufficient knowledge of the methods and procedures involved in handling the enlarged variety of ammunition now is use" (p. 267). Also, storage charts were made available to and studied by trainees prior their going on to fieldwork (p. 273). Such charts emphasize visual literacies.

Training—Manuals/SOPs

Visual and experiential modes of presentation were emphasized in training the lineworkers, and they were important to training of supervisors. However, there are manuals and standard operating procedures (SOPs) available for both lineworkers and supervisors. These require different kinds of literacy skills than the visual, experiential, and aural practices.

Generally, among the pre-accident manuals and SOPs that I observed, many were print-linguistic-heavy, incorporating few graphics. Figures 5.4–5.6 are three consecutive pages from the Ordnance Inspection Manual of 1942. This document is 280 pages long, and it would be read by inspectors and supervisors. As the examples show, it is print-linguistic text-heavy.

In addition to the inspection manual, a primary manual used by supervisors and workers also integrates much print-linguistic text. Figure 5.7 illustrates the only page from the Manual of Safe Practices (1941) directed at employees who handled explosives.

This manual was provided to all employees, and different sections of it, acknowledging particular information for workers who worked in certain positions, were color-coded for each position. These manuals use a considerable amount of print-linguistic text, suggesting that employees also needed to have related literacy skills.

Post-Accident Training—Manuals/SOPs

There does not appear to be a change in the visual, aural, and experiential approaches used in training after the accident, largely because the TWI program was still in place. It was the primary model of training throughout WWII. However, there is a dramatic difference in the print-linguistic materials used with that training after the accident. Figure 5.8 is a page from a manual used by supervisors that was published in 1945, two years after the accident.

PROCESSING OF AMMONIUM NITRATE AT RAVENNA ORDNANCE

PLANT

Receiving and Unloading: When shipped, cars are
equipped with a half length "dip pipe" installed.
On arrival, this pipe is immediately removed and
the full length "dip pipe" (which is strapped to
the under frame of the car) is substituted.

When the half length dip pipe is removed from
the car, a sample of the clear liquor is taken
through the man hole for analysis by the inspector.
He should promptly test for the presence of acid
impurities with litmus paper. If the solution is
alkaline, the unloading may proceed and the remainder of the sample is then taken to the laboratory
for "heat test." The "heat test" was developed at
Ravenna Ordnance Plant as a check on any act of
sabotage in transit and is not a part of the specification. The principle is that pure NH_4NO_3 free
from organic substance or acids will sublime with
a clear fume leaving no residue.

The method of inserting the full length dip
pipe is as follows: Remove the short pipe by un-
bolting the flange connection (b) and insert the
long one as far as it will go without forcing,
attach a steam hose and turn the steam on in the
pipe slowly. This will dissolve the crystals
around the pipe and allow it to settle into position for bolting securely to the flange.

The "dip-pipe" is then connected to the liquor
unloading line. The air hose is connected to its
fixture and with all other valves and the manhole
tightly closed, air pressure to the extent of 25
pounds is applied and thus the supernatant liquor
is blown directly into the storage tanks.

The air pressure gauge reading is checked at
regular intervals during unloading to see that the
air reducing valve is operating properly. The air
pressure on the inside of the tank car is not to

-18-

Figure 5.4. Page from manual used by workers.
Document courtesy of the U.S. Army.

exceed 25 pounds per square inch, a safety factor
established by tank car interests.

   The approximate time for this first blow is
one hour. At the end of this time the "man hole
cover" is then turned back on its hinge and the
agitator **pipes** are inserted lengthwise of the car,
reaching about 2/3 of the way to the ends of the
car. These pipes are of stainless steel, with
steam holes so placed that they open to the top
of the tank car--to prevent direct steam on the
sides and bottom of the car. When the agitators
are inserted, it is not necessary to disconnect the
"dip pipe" and air lines. This prevents loss of
time.

   The agitator pipes are connected to a steam
line of reduced pressure, and a vigorous flow of
steam is maintained until the ammonium nitrate
crystals are completely dissolved. When the bottom
ends of the car are hot to the touch, one may safely assume that the contents are entirely in solution
and the final blow off is made.

   In order to make the final blow off of the
hot liquid into storage the agitator pipes are
removed, the man hole cover sealed, and the air
pressure applied.

   When the liquor has been removed the manhole
cover is opened and the car checked for the complete
removal of solids and liquids. All unloading equipment is then washed down with clear water and all
caps and covers are replaced on the dome of the
tank car.

DIAGRAM OF UNLOADING OPERATION

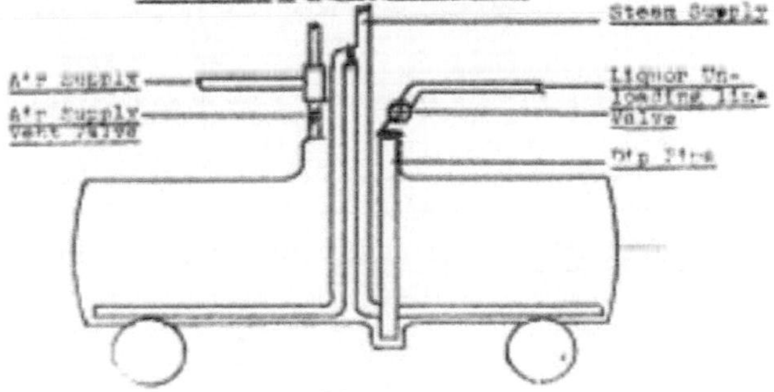

Steam Supply

AIR SUPPLY

Air Supply
Vent Valve

Liquor Un-
loading Line
Valve

Dip Pipe

Figure 5.5. Second page in sequence.
Document courtesy of the U.S. Army.

Storage: The Savanna Ordnance Plant has storage facilities for about 32 tank cars consisting of 4 storage tanks of 85,000 gallons each. This neutral liquor storage room is equipped for the maintenance of a temperature of about 56° C. by the use of a very large thermostatically controlled unit blower and an elaborate system of ventilation ducts. However, in warm summer weather the system is not used at all times as it is possible to keep the liquor in complete solution by the intermittent application of steam through the tank agitator pipes. These agitator pipes radiate from the center of each tank at 6 points and are about 3 inches from the bottom of the tank. During extremely cold weather these agitator pipes may be used as booster heaters.

Each tank is equipped with a device for measuring the volume of the liquor but these tank-meters are not satisfactory due to the corrosive action of the ammonium nitrate. The method used at S. A. Y. is to lower a float attached to a tape into the storage tank and by difference, from a given mark, the height of neutral liquor in the tank is obtained. A graph has been drawn at the S. A. Y. laboratory so that if the specific gravity of the neutral liquor at 56° C. is known, the percent of ammonium nitrate can be read directly. The specific gravity is determined by using hydrometers with a range of 1.27 to 1.45 and which are accurate to .01. When using the formula, (Sp. Gr. of Neutral Liquor x 61.5) x (706.9 x height of tank) x % of ammonium nitrate = dry weight of ammonium nitrate) to check on receipts from the vendor, there has been a disagreement of less than 1 percent. In the above formula 61.5 is the weight in pounds of 1 cubic foot of water at 56° C; 706.9 is the calculated area of each tank in square feet.

The Neutral Liquor Circulating System: In starting the flow of liquor, from the tanks to the pans, after a week-end shut down or after a break down it is essential that the pipes be heated in order to dissolve any small pockets or films of nitrates that may have formed along the lines. This is done by using hot water, at a temperature of about 150° F, which is kept in a 6,000 gallon tank located in the pump room. This water is kept ready at all times, as it is used to wash out the lines whenever it is necessary to stop the flow of liquor.

Figure 5.6. Third page in sequence.
Document courtesy of the U.S. Army.

33

GENERAL SAFETY RULES COVERING
ALL EXPLOSIVE OPERATIONS

1. In case of fire or explosion go to nearest shelter as quickly as possible, unless you are a member of the Fire Brigade—in which case you will go to your assigned station.

2. Do not allow any unauthorized person to operate your equipment. When you need relief call your Foreman.

3. Do not operate any defective or unsafe equipment; shut down and report to your Foreman. This includes hot bearings, loose glands, presence of static electricity, etc.

4. Do not attempt to repair, alter or adjust any of your equipment but report the condition to your Foreman. Such work must be done by the specified repair man.

5. Keep building, floor and equipment clean at all times.

6. Keep passage-ways and exits clear at all times. See that exits open freely. Remove ice, snow, or any obstructions before starting operation.

7. Do not sweep any explosives out of the building. Dispose of such materials by placing in the proper can.

8. Always use non-sparking tools unless otherwise specified.

9. Do not operate equipment until you have been properly instructed.

10. Unlock all doors of a building before starting work in the building.

11. See that all ventilation fans are running while you are in the building.

12. Always use the proper type of air hose with a grounded metal nozzle, and avoid any whipping of the hose.

13. No lunches or parcels of any kind can be carried into an operating building.

Safety Takes Cooperation

Figure 5.7. Page from 1941 Safety Manual.
Document courtesy of the U.S. Army.

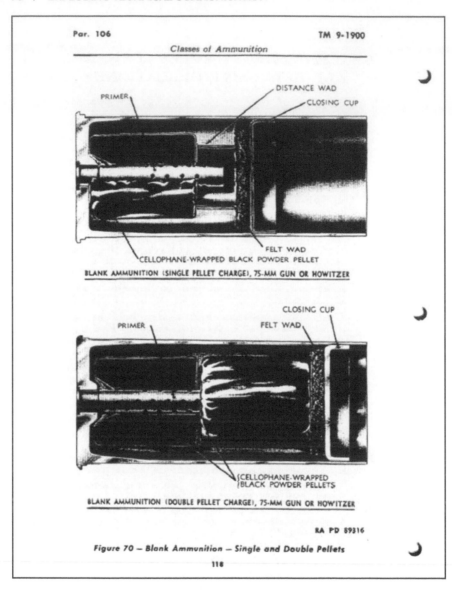

Figure 5.8. Page from Supervisor Manual (1945).
Document courtesy of the U.S. Army.

The page integrates images of two different kinds of ammunition, helping the reader visualize the particular item being described by text. Figure 5.9 shows a two-page spread taken from a bomb manufacturing SOP, also published in 1945.

The spread provides photographs of the particular step, and these pictures are positioned directly above the print-linguistic text. Mayer (2001) encourages providing images and textual information together so that each reinforces each other. Mayer's multimodal principle is that people learn better when pictures and words are integrated into an instructional message than when only words are used (p. 63). When only words are used, people may attempt to "build a visual model," but they may not attempt to do so. If a picture is provided, people can make the visual connection more readily. Note also the print-linguistic text regarding safety positioned to the left, among the first items one would read when reading a traditional print-linguistic text from left to right.

Finally, even the print-linguistic text in the supervisors' manual calls attention to safety. This includes the handling of fuses and issues with dropping packages (see Figure 5.10). Not only is the information provided explicitly, but it is italicized. The visual feature of using italics to highlight the text calls the reader's attention to it.

In addition to using print-linguistic literacies in training after the accident, more effort was made to integrate visual attributes into print materials to appeal to the visual literacies of workers and to reinforce the visual aspects of the experiential and aural training. The highlighting of certain text related to safety and positioning of it in certain places on a page contribute visual attributes to the print-linguistic text too, reinforcing the message. These are noticeable differences between the print-linguistic materials available to workers before the accident and those available after the accident.

In addition to integrating more graphics and text information in visually sensitive ways, readability tests show changes in how print-linguistic attributes of the SOPs and manuals used in training were prepared. Flesch-Kincaid scores of text decrease over time. This suggests a progressive understanding of the need to modify print-linguistic texts to accommodate low literacy levels. In particular, grade-level readability scores for manuals/SOPs seemed to drop from 1942, before the accident, to 1953, after the accident. Table 5.1 shows this comparison, including publication date in parentheses.

Lineworkers and supervisors would have been readers of the Inspection Manual and the Planning Manual. Supervisors would be the only audience for the Ammunition, General. The 1942 manual, which predates the accident, has the highest grade-level requirement, and the 1953 manual has the lowest. This represents a change over time in understanding of the print-linguistic literacy skills of the employees. It also represents an effort to sponsor print-linguistic skills through adjusting the text for a lower print-linguistic literacy level. The grade-level skill required to read and understand the information goes from almost a 12th-grade level in 1942 down to less than a 7th-grade level by 1953.

Stork" for transporting Tritonal to pellet-making bay.

Loading pellets into "hat boxes" for transfer to tail pour.

Breaking into pellets.

OPERATION VI-B - PELLET MANUFACTURE

PERSONNEL PER SHIFT	EQUIPMENT, TOOLS AND GAGES	MATERIAL
Production Standard Crew	Transfer tub or "stork", cooling pans, breaking table with screen, mallets, pellet boxes, dust collection box.	Tritonal form V (slightly thinner than regular bomb pour).

SAFETY	QUALITY	METHOD
1. Maintain good housekeeping.	1. Check samples of pellets for density and absence of cavitation.	1. Receive Tritonal from Operation V in transfer tubs or "storks".
2. Observe general safety rules for handling explosives.	2. Check for excessive segregation.	2. Pour into pans, not thicker than one inch.
3. Avoid splashes and overflows.	3. See that pellets of proper size are produced.	3. Allow to stand undisturbed until completely solidified.
4. Operators should wear protective equipment.	4. Make certain pellets are properly screened, removing excess dust and fines.	4. Remove from pans and place on breaking table. Break into proper sized pellets. The size of the pellets will vary with the size of the bomb being manufactured.
5. Use non-sparking tools.		5. Screen dust from pellets and place pellets in boxes for future use.
		6. Collect fines for use in Dopp kettle (Operation V-5) or in Operation VII-8.

Figure 5.9. Two-page spread of bomb SOP guide.

Source: From Army Service Forces, Office of the Chief of Ordnance, Office of the Field Director of Ammunition Plants *Standard Practice Manual 250 LB. Bomb, G.P., AN-M57A1 500 LB. Bomb, G.P., AN-M64A1 1000 LB. Bomb, G.P., AN-M65A1 2000 LB. Bomb, G.P., AN-M66A1 Tritonal Loaded.* Document courtesy of the U.S. Army.

TM 9-1900   Par. 219

Care, Handling, and Preservation

the round from a hot tube within 45 seconds after the original misfire, water should be played on the barrel until it is cool. The safest time to remove a misfired round of fixed ammunition is between 30 and 45 seconds after its occurrence.

(2) SEPARATE-LOADING AMMUNITION.

(a) Two attempts will be made to fire the primer before it may be removed. If the primer is heard to fire, a minimum of 60 seconds will be allowed before the breech may be opened and the faulty charge removed. The faulty charge must be stored separately from other charges.

(b) If the primer is not heard to fire, two more attempts to fire will be made. Then proceed as follows:

1. If the primer can be removed by a person standing clear of the path of recoil, after 2 minutes have elapsed, the primer may be removed and a new one inserted. If the second primer fails, 10 minutes should be allowed to pass and then the breech may be opened.

2. If the primer cannot be removed safely as described above, no attempt will be made to open the breech or replace the primer for 10 minutes.

(c) Misfire primers should be handled carefully and disposed of quickly, owing to the chance of a primer hangfire. Further information will be found in AR 750-10 and the Technical Manuals and Field Manuals pertaining to the piece.

g. Fuzes.

(1) *Extreme care must be taken in handling and assembling fuzes to shell or bombs. All fuzes must be treated as delicate mechanisms. The forces which arm a fuze on firing a weapon can be simulated by rolling or dropping, and a fuze so armed may be functioned by the impact of a blow or by dropping.*

(2) In the assembly of fuzes and projectiles, the fuze body, threads, adapter, and fuze cavity must be inspected to insure that grit, grease, or other foreign material is not present. This is necessary for proper seating of the fuze without the use of excessive force. Cleaning of the fuze cavity should be accomplished with a piece of cloth and a small stick which can be inserted into the cavity. Fuze-hole lifting plugs should not be removed except for inspection or when the fuze is about to be inserted.

(3) When ammunition or projectiles are issued fuzed, no attempt will be made to remove the fuzes without specific authority and instructions from the Chief of Ordnance.

(4) Fuzes will not be altered. Any attempt to alter or disassemble fuzes in the field is dangerous and is prohibited except under specific direction of the Chief of Ordnance. The only authorized assembling or disassembly operations are screwing the fuze into the

303

Figure 5.10. Page from Supervisor Manual (1945).
Document courtesy of the U.S. Army.

Table 5.1. Comparison of Document Readability

Document	Page number sampled	Flesch-Kincaid score (Grade-level)
Ordnance Inspection Manual (1942)	18	11.49
Ordnance Inspection Manual (1942)	33	11.7
Ammunition, General (1945)	117	8.99
Ammunition, General (1945)	210	8.91
Mobilization Planning Manual for 90 mm Complete Rounds (1953)	3	5.87
Mobilization Planning Manual for 90 mm Complete Rounds (1953)	4	6.46

Indeed, the material published shortly after the accident shows a drop in grade-level expectation to less than a 9th-grade level. As the grade-level reading skill was adjusted downward, employees may have been able to better understand instructions and information provided in print-linguistic form. Authors of these texts may have received feedback reflecting this understanding and adjusted their writing accordingly.

Several documents illustrate a shift in the way information is presented in print training materials. These shifts indicate a change in sponsorship dynamics, recognizing that more visual information in them may encourage workers to review them and understand them better than using only print-linguistic text. Also, the readability test results on the manuals suggest writers adjusted over time to facilitate easier reading of the materials for workers.

While these print materials exist, however, it is also important to ascertain how people used these documents. As Purcell-Gates (1995) notes, the presence of books in a home does not mean that they are read. Similarly, workers in a manufacturing workplace use print materials to varying degrees. Hutchins (1995), for example, notes that print-linguistic instructions tend to act as a guide more than as a standard for all settings in which a given task may occur.

Employees of the Arsenal did not take the manuals home and read them carefully. The people whom I interviewed who had worked at the Arsenal indicated that they never brought workplace documents home or were even permitted to do so. In response to my question about what employees may have brought home from the Arsenal or talked about of work at home, Steve, an interview subject who worked at the Arsenal in the 1960s and 1970s, acknowledges that there were limitations on what employees could bring home or talk about:

> Steve: No, no, pretty much. Pretty much in those days we had security clearances and the protocol in those days was to sort of leave work at work. Don't come home and talk about shipments or what you were doing.

Neither can one assume that lineworkers read much to prepare for a job at work. Few of those interviewed recalled having to read much material, and they emphasized the on-the-job training experience, which involved visual and aural literacies. While some people acknowledged that manuals were available, they also stated that these were used more as reference materials than as primary instructional documents.

Hutchins (1995), who reports about operations in a nuclear submarine, documents field practices that do not follow training or documented instructions and how such practice becomes accepted behavior. He calls this practice "situated cognition." Hutchins explains that a set of documented procedures acts as a "meditational device" between the task and the reader, however, he also observes that learning can "be mediated by so many different kinds of structures" (p. 291). These structures include visual/experiential modes. Hutchins characterizes written instructions as guidelines. Amerine and Bilmes (1990) point out that written instructions serve as a set of guidelines, and authors of instructions often omit information that they feel the reader may be able to infer. Hutchins, further, observes that reading instructions and performing the steps represent different mediating structures for the learner; the act of reading a single written step in a procedure involves understanding "what the step says, what the step means, and the actions in the task world that carry out the step" (1995, p. 301). Hutchins explains that written instructions act as a guideline around which actual performance in doing the task may be negotiated because of the potential to infer actions from written instructions that may include limited information.

Again, Arsenal workers were not familiar, generally, with munitions prior to their work at the Arsenal, however, Hutchins' (1995) subjects were trained nuclear submarine specialists. As specialists, they would have specialized knowledge of particular operations and would not need highly prescribed instructions in training materials. Potential hazards of allowing readers to infer actions in the absence of explicit information is greater for unskilled personnel like those who worked at the Arsenal than it is for those who have the degree of training like those operating a nuclear submarine. So Arsenal workers should have been given more explicit information, as there was in texts published after the explosion.

Training practices that predate the accident emphasize literacies other than print-linguistic; they draw the focus of practice to visual, experiential, and aural literacies. The TWI program was used at arsenals throughout the country during WWII, and documents associated with that program show that it purposefully emphasized these literacies because of employees' backgrounds. However, some of the workers who worked there after the accident acknowledged being encouraged to read the safety manual as part of their training and subsequent to it.

David, who worked there after the explosion, acknowledges an emphasis on safety protocol among lineworkers:

> Me: What kind of training was there associated with that line work that you remember?
>
> David: I had some . . . there wasn't any training outside of safety. They had like a change house and on the actual line as far as pouring the melt. I didn't do this, but, on the line where they melt the TNT; that is not the name of it, but that's what it is; okay . . . and the . . . as for safety, we were not allowed to have any matches or, if you . . . like . . . there had been a change house a short distance away probably about . . . probably about 300 feet, 400 feet, something like that; and that's where they allowed the people to smoke. No lighters and no matches were allowed in there at all. But they had a small lighter that was connected to electrical. And if you tilted it a little bit, it would light for the people who wished to smoke there.

Don, who also worked there after the accident, also articulated the existence of a manual that emphasized safety:

> Don: I was almost guaranteed a job before I ever got out there. It was in safety and security, which was a field I knew nothing about. But nonetheless there was a field manual that you carried with you and that manual had been written in blood; and, so, you didn't vary too much from what was in the field manual.

Don recalls, specifically, that workers were strongly encouraged to link their practices with information from the manual. This indicates a closer relationship between print-linguistic practices and actual physical practice.

David points out that training emphasized safety information, and Don acknowledges the existence of a safety manual that was treated as very important. While Don's position is directly related to safety, the acknowledgement of the manual and its importance suggests that workers were encouraged to use the manuals. The separation of modes contributed to the accident and that combining modes of representation in a way that reinforces each other, the information conveyed is better remembered and practiced.

LITERACIES RELATED TO TRAINING: IMPLICATIONS FOR SPONSORSHIP

Training of workers and supervisors—pre-accident and postaccident—emphasized visual, aural, and experiential literacy practices in order to help people who were not experienced with war industrial work to learn appropriate skills in an efficient manner, in accordance with the TWI Program. Considering the protocol articulated in the accident report, workers were trained to unload bombs as a two-man operation that was a safe way to handle the boxes. However, in practice, this was modified to make the operation go more quickly. While not indicated either as acceptable or unacceptable practice in the written procedures for unloading bombs, the routine procedure of dropping boxes and walking them evidently had become an accepted practice to accelerate unloading of bombs. Stratton (1943) acknowledges in the investigation report that such modification "was common practice" (p. 7). It may have become accepted practice because of workers' prior experiences with fuse tolerances that prevented safety-related consequences. The modification to a one-man unloading operation from the safer two-man operation may have been encouraged in practice, not related to training.

Long-Range Changes in Manuals

An interesting observation among the manuals designed with lineworkers and inspectors as primary readers is a trend over a period of time in the use of graphics. While graphics in the 1945 bomb SOP are positioned to be immediately available to readers as they view print-linguistic text, manuals published in the 1950s position graphics, mostly photographs, as appendices. Such separation and placement of graphics suggests primacy of print-linguistic text over visual representations.

Such differences among the various print materials signal a shift from visual/ multimodal compositions to more print-linguistic compositions. This suggests a shift in the literacy sponsorship toward esteeming and facilitating more print-linguistic literacies than the visual literacies facilitated in the 1940s.

As mentioned above, the government and APCO were the operators of the Arsenal during WWII, when there was a shift from print-linguistic-heavy manuals to manuals with more balance between print-linguistic text and graphics. This pattern of balancing print-linguistic text with images suggests a shift in the way the agents sponsored literacy within the ecology of the workplace— the Arsenal. As they recognized the need to facilitate communication with low-literacy workers, they integrated more balance to encourage workers to look at the materials and to read the print-linguistic text accompanying the images. Further, the adjusting of grade-level readability of print-linguistic text in the manuals suggests that the agents were trying to encourage workers to read the manuals by making the information more understandable. However, this balance and use of graphics shifted in materials published in the 1950s.

Vulcan Tires was the contractor and primary sponsoring agent in the 1950s, when photographs were not used in the newsletter and when photographs and graphics were placed in appendices in manuals and SOPs. The changes could be related to the changing workforce. As stated above, those who worked at the Arsenal during WWII came from various literacy backgrounds and had no experience with war-industry production. They would benefit from having images to refer to as they learned to manufacture munitions. As skilled veterans returned from war, they could take on those jobs without needing much training or as many images to help them understand their tasks. Also, most people who worked at the Arsenal after WWII had been employed there during the war. These workers would not need the kind of training associated with the TWI Program. None of the historical summaries of operations at the Arsenal from the 1950s indicate anything about the training that occurred there.

Changes in Training

Again, no accidents of the like occurred after this event. As indicated earlier, Stratton (1943), the author of the investigation report, emphasized acknowledgment of the need for caution with the fuses, and that acknowledgment is evident in subsequent manuals. Additionally, a few people who worked at the Arsenal also recalled an emphasis on safety. No other accidents involving such explosions occurred subsequent to this accident, suggesting that these safety precautions were clearly articulated in writing and reinforced several ways.

CONCLUSION

The training program emphasized visual, aural, and experiential practices. While manuals and SOPs existed, they were not used much. Any print materials that were used before the accident required advanced reading skills that many lineworkers and depot workers likely did not have. Consequently, workers would not have been able to read that material. Further, those print materials did not include specific information about safe handling of equipment. Any such safety precautions would have been given to workers as they observed demonstrations and practiced in the field or on the line.

While training continued to emphasize nonprint literacies after the accident, print materials changed after the accident to include more visuals that were placed within the text itself, and the materials were easier to read because the grade-level requirement to read them fell considerably. This enabled people to read those materials. The visuals and print-linguistic information reinforced any information given orally and through demonstrations. Also, it is clear that safety precautions articulated in the print materials were reinforced orally.

Finally, print materials used in training changed in the 1950s to position visuals after print-linguistic text information. This suggests a shift in the sponsorship of visual literacy toward esteeming print-linguistic literacies over visual literacies. This would be consistent with the general valuing of such literacies throughout the country as demonstrated by the G.I. Bill's efforts to encourage college attendance and training, which generally included print-linguistic skills development.

The practices leading up to the accident and the actions taken after it show that it is important to connect print-linguistic practices with other literacy practices to ensure that information is reinforced in multiple ways. The next chapter describes findings regarding other literacy practices at the Arsenal that show how differences in organizational communication practices across levels within the organization's hierarchy contributed to the accident.

CHAPTER 6

Other Literacies at Work

In the previous chapter, I showed how technical communication associated especially with training practices and disconnections between the ways in which information was provided orally and in print documents contributed to the accident. I also showed how changes in the materials used in training toward making connections across materials and practices clearer helped to ensure that a similar accident did not occur again. In this chapter, I show how other literacy practices at the Arsenal, including managerial communication, contributed to the accident and how some of those practices changed subsequent to the accident and after WWII. Again, a memo acknowledging the defect in the fuse existed, and it is clear that administrators knew of the defect, encouraging workers to be careful with the shipment. However, differences in literacy practices across levels within the organization may have contributed to the accident in a number of ways. Northey (1990) observed different kinds of practices going on at different levels in accounting firms, suggesting that those in higher management positions were expected to have better print-linguistic literacies than those at lower levels of management. Similar dynamics occur in this study as well, and I alluded to this phenomenon before (Remley, 2009). Further, a notable change in print materials occurred in the 1950s to suggest a return to favoring print-linguistic literacies, however, a number of variables contributed to that shift. I detail such practices and differences here.

While training tended to emphasize visual and experiential literacy practices, in addition to manuals, employees also had access to newsletters and forms. In addition to reading the same newsletters and most of the forms, managers and administrators composed and read various routine and special reports. Newsletters combined print-linguistic and visual texts while reports tended to integrate more print-linguistic text while integrating some visuals. In this chapter, I respond to particular questions associated with these literacy practices across levels in the Arsenal's organization. In addition to identifying specific practices, I address the question, What intersections between these practices exist? Inferences from answers to these questions can address the following questions: What groups may have benefited from this sponsorship? and What groups may have been negatively affected by this sponsorship?

Those who worked at the Arsenal reported very few writing practices at work. This suggests that most interviewed did not have an administrative or managerial position. Generally, other than administrators, only those who were employed as secretaries did any kind of writing on a regular basis, and that was transcribing what was dictated to them. Table 6.1 shows a breakdown of literacy practices among interview participants who worked at the Arsenal.

Table 6.1. Literate Practices at the Arsenal

Participant	Position	Reading practices reported	Writing practices reported
David (self)	Supervisor	Newsletter, manual correspondence, reports	None
Sarah (self)	Secretarial	Correspondence, reports, manuals	Correspondence, reports, manuals; transcribing
Sandy (self)	Secretarial	None	Correspondence, manuals, reports: transcribing
Roger (self)	Line/labor and supervisor	Reports, correspondence, manuals, newsletter	None
Susan (relative)	Line/labor	None	None
Steve (self and parent)	Supervisor (both)	Reports/manuals	Reports
Claire (2 relatives: husband; brother)	Line/labor	Manual	None
Carol (2 relatives: parents)	Construction; secretarial	Reports, manuals, correspondence	None
Jane (relative)	Line/labor	Manual	None
Don (self)	Construction	Manual	None

This transcription practice is evidenced with the following interview dialogue:

> Me: What kind of reading are you doing; reports or is it manuals, what?
>
> Sarah: It's actually reports. And the person I work for was the head of the employment services division. And it was mostly letters in and doing letters for him and things like that.
>
> Me: So, it was taking dictation and then writing the letters for him?
>
> S: Yes

Those who supervised also wrote periodic reports. A practice acknowledged in the following interview:

> Me: Okay, you mentioned that your father worked at the Arsenal and that he participated in the construction of it. What did he do at the Arsenal when he started there?
>
> Steve: He was in supervision and supervision in various titles for a number of years, and he retired in 1982 after 42 years at Boomtown Arsenal.
>
> Me: Did he ever talk about the kind of work he did there? Any kind of reading or writing associated with that work?
>
> Steve: He did. He talked about reports. Of course, when you're with the government; he was in the government for 11 years and then he worked for the contractor. The Boomtown Arsenal became a contractor operated facility in 1952 and those who stayed on with the contractor, they had that opportunity to do that.

As acknowledged above, literacy materials used by different audiences may reflect a certain understanding of the primary audience's literacy skills. Low-literacy people or those learning a new skill tend to favor using visual and oral materials and experience rather than print-linguistic materials (Mayer, 2001).

However, the relationship between alphanumeric, print-linguistic literacies and visual literacies also depends on the rhetorical nature of the material involved and the audience. Instructions tend to require extensive use of visuals, while reports tend to require fewer visuals and different kinds of graphics than those typical of instructions (Markel, 2010; Oliu et al., 2010). Reports read principally by administrators and supervisors had more print-linguistic content than other forms of representation. As indicated above, manuals and Standard Operating Procedures (SOPs), which were read more by lineworkers than anyone else, included many more graphics than reports did. Also, the training lineworkers and supervisors received emphasized visual and experiential literacy practices.

WHAT IS THE RELATIONSHIP BETWEEN
PRINT-LINGUISTIC TEXT AND GRAPHICS?

Generally, a large amount of graphics used in a given document suggests an appeal to readers with visual literacies, while a limited use of graphics with an emphasis on print-linguistic text would appeal to readers with print-linguistic literacy. Lineworkers and inspectors came from varied literacy backgrounds. Many were unskilled in industrial work and more familiar with farmwork. They were the primary readers of the manuals and SOPs. Therefore, authors of these texts applied a greater use of graphics to the materials for those employees than print-linguistic text to facilitate meaning-making. Such visuals included samples of templated forms inspectors would complete, including the one displayed in Figure 6.1.

Such a form, which required reading and writing, minimized print-linguistic literacy demands by providing labels for most of the textual material needed and organizing it visually. This suggests lower literacy requirements for workers who completed it since most of the information requested was provided or prescribed, reducing the amount of work they had to do for the writing task.

RHETORICAL PURPOSE OF DOCUMENT AND
RELATED READERSHIPS

Generally, there appears to be several differences in the use of graphics across different kinds of documents relative to their general readership and literacy skills associated with those readers. However, these differences can also be attributed to the purpose of the document. Much scholarship finds that certain kinds of graphics are more useful than others for particular rhetorical purposes: generally, for example, photographs closely approximate physical objects they represent, making them useful in instructions, while tables effectively represent numeric data for technical or managerial reports in which such information is presented (Gurak & Lannon, 2007; Helmers, 2006; Kolin, 2009; Murray, 2009).

Further, as acknowledged previously, Mayer (2001) and others support the use of multiple modes in education and training materials because the different modalities can reinforce each other, or certain modes may appeal more to certain readers than to others, and including both will facilitate learning for different kinds of learners. Murray (2009) also finds the value of ambiguity in providing print-linguistic text alongside of related images in multimodal compositions (p. 32). As acknowledged above, material read by lineworkers—newsletters and manuals—tended to integrate a lot of photographs, while supervisors' material included more diagrams, and executives' materials integrated more tables.

Table 6.2 shows the mean percentages of the use of graphics relative to different kinds of documents.

Generally, the historical summaries and building specifications integrated the most print-linguistic material by percentage of total pages.

(SPECIMEN)

PRODUCTION REPORT L. L. #3

Shift No. _____   Date _____

Type of Bomb _____   Mark No. _____

No. Shipped _____   Ex. O. No. _____

No. Poured _____   Amm. Lot No. _____

Car Nos. _____   Data Card No. _____

_____   Body Lot No. _____

_____   Fuze Seat Liner Lot No. _____

_____   Fuze Seat Liner Type _____

_____   T.N.T. Let No. _____

                                   NH_4NO_3 Lot No. _____

GAGE USUAGE   Amatol _____
Type _____P-_____Times_____ Aux. Booster Lot No. _____

_____ _____ _____   Adaptor Booster Lot No. _____

_____ _____ _____   TIME REPORT

_____ _____ _____   Name_____Hrs.____Bldg._____

_____ _____ _____   _____   ____ ____

_____ _____ _____   _____   ____ ____

T.N.T. Temp. _____   _____   ____ ____
NH_4NO_3 Temp. _____   _____   ____ ____
Amatol Temp. _____   _____   ____ ____
% Rel Humidity _____Temp._____   _____   ____ ____
NH_4NO_3 Sample No. _____
Density Tests. Shift No. Poured_____Date_____Result_____

(This form is filled out to show the production
of each 8 hour shift.)

Rejents:

Remarks:

-171-

Figure 6.1. Inspection form used by line workers.
Document courtesy of the U.S. Army.

Table 6.3 shows the percentage breakdown of certain kinds of graphics relative to documents.

Photographs and diagrams ware used heavily in manuals, while tables ware used more in historical summaries, special reports, and building specifications.

Table 6.4 shows the percentage breakdown pertaining to size of graphics by document.

Table 6.2. Use of Graphics

Document type	Text-only	Image-only	Combo
Manual or SOP			
Mean	57.4300	38.1483	2.2600
N	6	6	6
Newsletter			
Mean	3.8281	6.1719	90.0000
N	16	16	16
Historical summary			
Mean	81.3264	9.6709	9.0009
n	11	11	11
Special report			
Mean	55.1860	40.4530	2.0740
n	5	5	5
Building specifications			
Mean	74.5420	22.9620	2.4940
n	5	5	5
Total	45.3270	17.4673	36.6374
N	43	43	43

Of particular interest is the use of graphics, which were more than 90% of a given page in manuals and SOPs. Generally, these were also photographs showing people performing a specific operation. This approximated to a large degree the reader's actual practice of observing someone performing the task.

Modal differences across documents also pertained to different primary readers of those documents. Table 6.5 shows the breakdown of use of graphics generally for different readerships.

Documents read primarily by lineworkers had more pages with visuals than did material read by any other readers. Documents read primarily by administrators used more print-linguistic pages.

Table 6.6 shows the percentage breakdown for certain types of graphics for different readerships.

As mentioned in Chapter 5, readers tend to favor photographs than other graphics because photographs provide a representation of the actual object as it appears to the reader. This is important when training or instruction is

Table 6.3. Mean Use of Graphics

Document type	Photo	Table	Chart	Form	Diagram	Line	Map	Drawing
Manual or SOP								
Mean	37.5867	8.7433	8.5183	9.4333	33.7767	.0000	.0000	1.9383
n	6	6	6	6	6	6	6	6
Newsletter								
Mean	53.7738	13.9519	.4463	.0000	.1894	.0000	.0000	31.6450
n	16	16	16	16	16	16	16	16
Historical summary								
Mean	.0000	82.1618	14.5191	.0000	.0000	3.8955	.0000	.0000
n	11	11	11	11	11	11	11	11
Special report								
Mean	12.3080	45.0000	10.4000	18.6660	17.8720	.0000	5.0000	.0000
n	5	5	5	5	5	5	5	5
Building specifications								
Mean	.0000	28.8880	2.2220	.0000	28.8880	.0000	.0000	.0000
n	5	5	5	5	5	5	5	5
Total								
Mean	26.6847	36.0212	6.5365	3.4867	10.2207	.9965	.5814	12.0453
N	43	43	43	43	43	43	43	43

Table 6.4. Size of Graphics

Document type	Less than 20%	20-33%	33-50%	50-67%	67-75%	75-90%	Over 90%
Manual or SOP							
Mean	3.3400	8.7033	4.3867	2.3833	.4767	.2917	80.4150
n	6	6	6	6	6	6	6
Newsletter							
Mean	60.5481	27.3306	10.4325	1.0450	.0000	.3762	.2606
n	16	16	16	16	16	16	16
Historical summary							
Mean	63.6782	12.5409	4.4327	1.2945	4.9191	1.8736	2.2591
n	11	11	11	11	11	11	11
Special report							
Mean	13.5340	11.6660	11.3340	.0000	4.6160	15.3840	43.6660
n	5	5	5	5	5	5	5
Building specifications							
Mean	25.1380	10.2780	1.2500	.0000	1.1100	.0000	22.2220
n	5	5	5	5	5	5	5
Total							
Mean	43.7821	17.1437	7.0912	1.0526	1.9907	2.4488	19.5570
N	43	43	43	43	43	43	43

Table 6.5. Graphics versus Readership

Reader	Text-only	Image-only	Combo
Line worker/Inspector			
Mean	59.1640	37.8280	.4140
N	5	5	5
Supervisor or Manager			
Mean	70.2450	25.7600	3.9933
N	6	6	6
Administration or Executive			
Mean	73.1575	19.2903	6.8363
n	16	16	16
All			
Mean	3.8281	6.1719	90.0000
N	16	16	16
N	43	43	43

involved. Lineworkers read manuals, SOPs, and newsletters, which integrated many photographs. Supervisors read other kinds of manuals, and administrators read reports, which tended to integrate fewer photographs and more tables and charts.

Again, certain documents were read by certain categories of employees; consequently, there was considerable correlation between the findings associated with types of documents and findings associated with readership. Lineworkers read manuals and SOPs, which integrated many large visuals, while supervisors and administrators read reports integrating smaller visuals. Larger graphics may have helped readers understand visual relationships represented in them, such as would be the case with a diagram or photograph in a manual showing a reader how to perform a given task.

Manuals and newsletters tended to have a considerably large number of photographs and diagrams, while specifications tended to use tables and charts more often. Drawings and photographs were common in newsletters. Also, manuals tended to use larger images—photographs, template forms, and diagrams—than routine reports and building specifications. Newsletters most often integrated graphics with text generally, while other documents tended to locate graphics both within text and as appendices. Building specifications tended not to have many graphics. This suggests that print materials that were used by workers generally integrated more graphics than those used by supervisors and

Table 6.6. Mean Breakdown for Graphics versus Readership

Reader	Photo	Table	Chart	Form	Diagram	Line	Map	Drawing
Line worker/Inspector								
Mean	44.8800	8.8060	9.5480	11.3200	23.1160	.0000	.0000	2.3260
n	5	5	5	5	5	5	5	5
Supervisor or Manager								
Mean	.1867	25.4783	2.4133	.0000	38.5867	.0000	.0000	.0000
n	6	6	6	6	6	6	6	6
Administration or Executive								
Mean	3.8463	70.5487	13.2319	5.8331	5.5850	2.6781	1.5625	.0000
n	16	16	16	16	16	16	16	16
All								
Mean	53.7738	13.9519	.4463	.0000	.1894	.0000	.0000	.0000
n	16	16	16	16	16	16	16	16
Total								
Mean	26.6847	36.0212	6.5365	3.4867	10.2207	.9965	.5814	12.0453
N	43	43	43	43	43	43	43	43

administrators. Also, print-linguistic texts used by workers for instructional purposes (manuals) used larger and more graphics than those used for informational purposes (newsletters).

As I mentioned above, there was also a historical trend toward moving graphics in manuals from within text to appendices, leaving related text as text-only pages. Placement of photographs within text in the manuals/SOPs published in the mid-1940s suggests a consideration of photographs as complementary to print-linguistic text associated with them, while placement of photographs in an appendix in manuals/SOPs published in the 1950s suggests a consideration of those photographs as supplemental or secondary material.

The Arsenal published a monthly newsletter for employees, and this tended to focus on social elements, though some articles discussed safety issues as well. Figure 6.2 shows a page from the May 1942 newsletter. Further, though they were used in newsletters often in the 1940s and 1950s, photographs disappear from newsletters by 1960. Compare Figure 6.2, reproduced with Figure 6.3, which shows an example of a page from the June 1960 issue of the newsletter.

The format of the newsletter changed from decade to decade such that there was more use of images, especially photographs, in the 1940s and 1950s than in the 1960 issues.

Sponsorship Implications

Clearly, the various modes of presentation used in training worked to reinforce the information conveyed in each, while also facilitating practice with the equipment before actually handling explosive materials. However, a separation in literacy skills, and consequently in modes of representation used in communication, evidently occurred as one rose to higher levels at the Arsenal and the Army generally. More alphanumeric, text-based writing and reading were emphasized because of the nature of the paperwork and reporting that occurred at those levels. This suggests an interesting correlation between literacy training, modes of presentation emphasized, and level of employment attainment in the workplace.

As evidenced above, as one attained higher levels of employment in the organization, the more alphanumeric, text-based forms of representation were emphasized. As a literacy sponsor, while facilitating training of workers to perform specific tasks to produce and store munitions, the Arsenal's operators restricted access to upper management positions by establishing a literacy hierarchy that limited the access one with low literacy has to upper management positions.

Administrators read more print-linguistic material while lineworkers were exposed to fewer print-linguistic materials; this separation seems to have contributed to the accidental explosion in that workers were "instructed" to be careful, while specific details of hazards associated with the fuse were articulated

Figure 6.2. Page from 1940 Newsletter.
Document courtesy of the U.S. Army.

Figure 6.3. Page from 1960 Newsletter.
Document courtesy of the U.S. Army.

to officials in a memo. That workers continued to use the usual routine of dropping the boxes and walking them in the igloo suggests that workers may have simply been told to be careful without having specific details necessary to understand *how* to be careful. Further consideration of literacy requirements relative to readability of the documents suggests that employees needed to have not only basic print-linguistic skills in order to advance but also literacy skills associated with a certain grade level. In the next section, I report on findings related to this separation of literacy practices and expectations thereof.

PRINT-LINGUISTIC LITERACY REQUIREMENTS ACROSS LEVELS: READABILITY TESTS

Another way to examine literacy differences that are embedded in the documents is to conduct readability tests on the print-linguistic text relative to the readership of particular documents. This can articulate a difference in expectations of readers and accommodations in language used, especially when examining comparable documents across different readerships. Manuals were provided, for example, to lineworkers and supervisors. If significant differences existed between the readability of these documents, there may have been a finding pertaining to differences in literacies across readers. This would have carried literacy implications associated with transferring information articulated between administrators and then to workers.

The Flesch-Kincaid test results found that there are differences in the mean grade level required to understand text across different types of documents and across different readerships. If the memo referred to in the explosion investigation report was written for a reader with certain grade-level literacy skills, the message would need to be modified before it was articulated to workers who may have had lower grade-level reading skills. I report these findings here because they suggest a literacy-grade-level requirement.

Administrators and managers needed to have completed at least half a year of college education in order to be able to read documents, such as reports, that they typically read (see Figures 6.1 and 6.2). Lineworkers and supervisors needed less than a 10th-grade education in order to read the materials they typically read, such as newsletters or manuals (Tables 6.7 and 6.8). This suggests that any information shared between administrators would have been difficult for workers to understand if the message was not first modified.

Table 6.7. Flesch-Kincaid Means According to Readerships

Reader	Mean	n
Line worker/Labor/Inspector	9.5033	9
Supervisor/Manager	8.9340	10
Administration/Manager	12.6343	14
All	9.8064	11
Total	10.4459	44

Table 6.8. Flesch-Kincaid Means According
to Document Type

Document type	Mean	n
Report: Routine	12.9145	11
Report: Special	9.7690	10
Manual/SOP	9.3333	12
Newsletter	9.8064	11
Total	10.4459	44

Differences in the Flesch-Kincaid scores across different documents signaled that the authors of these texts understood that readers would have different backgrounds. While it is important to understand what literacy expectations there were as evidenced in the documents themselves, it is also vital to ascertain how people used these documents.

Sponsorship Implications

While TWI emphasized visual and experiential literacy practices to mitigate low literacy levels and facilitate efficient training and production, Dinero (2005) acknowledges that the TWI program fell out of favor in the United States after WWII, largely because of three factors: there was no longer a central entity to supervise its implementation, the expansion of production facilities and equipment, and the lack of the necessity of war-related efficiencies due to the absence of serious international competition to motivate continued use (p. 14). Dinero acknowledges the shortsightedness of this attitude, observing that companies were quickly trying to shift production from war goods to consumer goods without considering how TWI could benefit them.

As stated previously, Vulcan Tires took over operations from the government by 1952, suggesting the potential for two different literacy sponsoring agents. Also, the Arsenal was still under government control in the early 1950s, as use of photographs and color in newsletters lessened. So placement of photographs in manual appendices in 1950 and using fewer photographs in newsletters at that time may be more a government-driven decision than a contractor-driven decision. The decision to place photographs in appendices of manuals may have been affected by the fact that many employees who worked at the Arsenal during the Korean War had stayed on after WWII. These employees needed little additional training or refresher reading. So sponsorship dynamics may have changed because of the workers' previous experiences with and knowledge of the operations and procedures. Alternatively, the use of fewer photographs in newsletters then may have been an economic decision, and the lack of photo-graphs at all in 1960s newsletters may have been a contractor-driven decision to reduce production expenses. Nevertheless, it reflects sponsorship toward favoring print-linguistic literacies.

Generally, visual and print-linguistic literacies seem to have been equally valued in the 1940s. While practices in the early part of the war seem to have esteemed print-linguistic literacies, the materials published in 1945 seem to have valued print-linguistic and visual literacies equally. This could have been due to the bomb accident and the need to integrate more graphics and more explicit information into training materials. Indeed, the text pertaining to safety in handling fuses identified in the previous chapter, published in the 1945 supervisors' manual, was likely a direct result of the accident.

The separation of practices across levels of the organization served to com-promise safety in the scope of the bomb accident. Print-linguistic modes were favored at higher levels of the organization while workers' low literacy levels were accommodated with visual, aural, and experiential practices. At some point in the process of moving the message from the administrative level (favoring print-linguistic literacies) to the workers (who favored visual, aural, and experi-ential literacies), the message specifying the defects and how they could be more careful in handling the boxes were lost.

Recall the passage of the report acknowledging that one of the fuse assemblies of the bombs involved was new and defective in its tolerances:

> f. The chief weakness in the fuze is in the pinion column which when coming out of adjustment generally permitted the safety blocks to fall out and the fuze to become armed. This data report in the form of a memorandum to O. E. Ralston dated January 6, 1943, indicating the defects in these fuzes, will be requested from Kingsbury Ordnance Plant. (p. 11)

The defect in the fuse's tolerance and that "over one million of these fuses have to be re-worked before they could be made reliable" indicates a problem in their manufacture that could have been promulgated through practices that included erroneous procedures. Also, the evidence of a memo (print-linguistic document) that predates the accident indicates that upper management would be aware of the defects and would need to articulate cautions to supervisors and workers in writing and/or orally. Evidence suggests this exchange of information was done orally, at least for the workers:

> h. . . . Helper No. 3070 working in the railroad car substantiated the previous testimony but indicated the boxes containing the bomb clusters were difficult to handle inasmuch as they were not provided with handles. He stated they were instructed to give special care to this shipment as it was the first shipment of this type of ammunition that had been received at the Depot. He detailed the procedure of unloading the cars as described in paragraph 5

above, making the statement that the boxes were permitted to slide down between the arms of the workers, landing on the end, and then were walked to the truck. (p. 11)

This passage indicates that workers were told to be careful, but they were not informed of specific precautions needed related to the fuse's tolerance sensitivity; they continued to use the routine procedure of rough handling associated with the munitions with which they were more familiar.

In the 1940 safety manual mentioned above, no information about precautions related to dropping or rough handling of boxes was provided in the section addressing workers employed in the explosives areas. Further, a 1941 Ordnance School Manual (the predecessor to the Ammunition, General) referred generally to the condition of projectile fuses as they were stored (pp. 55, 59). Figure 6.4

c. Safety features. (1) Firing pin support. The firing pin support maintains the firing pin at a safe distance from the detonator. It is identical to that of the point-detonating fuze, M48.

(2) The interrupter. The interrupter, while in the unarmed position, closes the passage leading to the booster, preventing superquick action of the fuze in the event that the superquick detonator functions prematurely.

(3) Cotter pin. The cotter pin supports the plunger during transportation, thereby preventing accidental shearing of the shear pins and firing of the concussion primer. It must be removed prior to firing.

(4) Fuze set at safe. When the fuze is set at safe, the metal between the ends of the time train in the upper ring covers the graduated-ring pellet, and the metal between the ends of the graduated-ring powder train covers the body pellet. Under these conditions one or both time rings can burn completely without igniting the base charge in the fuze.

(5) Vent closing discs. The vent closing discs prevent premature ignition of powder trains by chamber gases. The pressure created by the combustion gases, upon ignition of the powder trains, ruptures these discs, thereby providing vents (termed "exterior vents") for the gases generated as the burning of the time trains progress. These discs also serve to seal the powder trains against moisture.

(6) Safety disc. The safety disc is located at the ig-

-55-

Figure 6.4. Fuze safety information published in 1941 Supervisors' Manual. Document courtesy of the U.S. Army.

shows the specific text pertaining to safety of the fuses from that 1941 manual (item C. 4). That text only indicates that there is some tolerance when the fuse is in the "safe" setting position.

Though no print-linguistic text explicitly indicated particular precautions, the information above suggests that there was safety protocol provided orally to workers in the depot area and that the protocol included moving boxes as a two-man procedure.

Also suggested with the memo is that, while a print-linguistic memo articulating specific concerns was provided, the particular nature of those concerns may or may not have included issues associated with dropping the boxes or handling them roughly, beyond safe practices conveyed in the two-man operation's protocol. Also, if such precautions were provided in the memo, as the word of necessary precautions made its way to the workers, specific information about the nature of care was omitted. Workers knew to be cautious. However, as suggested by the continued "routine" procedure of workers dropping boxes, they did not know specifically of what to be cautious or how to exercise caution. Had specific information about the fuse's defect and precautions contained in the pre-accident dated memo been circulated to all depot workers, either in document form or orally, the explosion may have been avoided.

However, as the accident shows, there are material consequences, including life and death, embedded in decisions to esteem certain literacy practices over others. While a literacy hierarchy is generally recognized in professional communication scholarship because of the different kinds of communication occurring at different levels of an organization, the literacy hierarchy at the Arsenal seems to have contributed to the bomb explosion accident. Specific information that was conveyed in a print-linguistic mode to an administrator and that could have helped workers understand how to be careful never found its way to the workers. At some point, the information was lost, and it may have been due to the separation in modes of representation encouraged at the different levels of the organization. Presenting information using multiple modes can reinforce or provide an alternative means by which to present that information.

Further, literacy materials were modified subsequent to the accident to make more connections across modes of representation through different levels of the organization. However, in the 1950s, employees' familiarity with procedures and operations at the Arsenal as well as a desire for cost-effectiveness contributed to a shifting back to esteeming print-linguistic documents generally while devaluing the role of graphics. Such a shift represents a general sponsoring of print-linguistic literacy.

The various practices at the Arsenal seemed to esteem print-linguistic literacies; the practices in the community also reflect the esteeming of print-linguistic literacies, suggesting the influence the government and operators of the Arsenal had as sponsors within the local literacy ecology. I discuss these connections more in the next chapter.

CHAPTER 7

Literacy in the Community and Home

Many studies of literacy practice describe intersections between school and home practices and how the more the two practices are assimilated in school the better students perform. There are also many studies showing how higher education programs and training programs can prepare workers for the practices associated with certain kinds of jobs. However, this study considered the impact that the literacy practices at a particular workplace had on the community—the homes, schools, and community centers—in which it operated. This chapter provides insight into these practices as they occurred at home, at school, and in community organizations in the Fieldview community. Generally, members of the community saw that those in esteemed positions at the Arsenal were expected to have print-linguistic literacy skills that lineworkers and similar positions did not need. The government, through institutions it provided the community, also reinforced a favoring of print-linguistic literacies. This combination contributes to the general sponsorship of print-linguistic literacies within the community. While multimodal literacies were practiced and facilitated at the Arsenal, it was for a limited period, and print-linguistic skills were still esteemed.

As mentioned before, the Fieldview community was the primary location where the Arsenal was situated. Many employees resided in the temporary housing development built in Fieldview or in other houses. Residents of a community who experience a certain kind of literacy sponsorship will practice that form of literacy in various settings outside of the particular sponsoring institution. Consequently, in the study, I researched the following questions pertaining to potential sponsorship the Arsenal's operators and the government exercised in the Fieldview community:

What literacy-related institutions or programs did the government or operators fund?
What literacy practices occurred in school?
What literacy practices occurred at home?
What literacy practices occurred in the community?
What intersections between these practices existed?

Inferences from answers to these questions can address the following questions:

What groups may have benefited from this sponsorship?
What groups may have been negatively affected by this sponsorship?

LITERACY-RELATED INSTITUTIONS

The government constructed and supported the Arsenal as well as a library within the Arsenal's grounds, the community's library, and an elementary school in the housing development built for workers at the Arsenal. The government's funding of these institutions suggests a form of sponsorship, since these institutions were traditionally associated with developing and encouraging certain literacy activities. The roles of these institutions in the community are evidenced in the interviews pertaining to home, school, and community-related practices. Because of the connection they had to the government's sponsorship within the literacy ecology of the community outside of the Arsenal, I provide some basic information about the construction of these institutions in this section.

Libraries

The Arsenal had a library on its grounds, which made a variety of books available to workers as well as children of those who lived on the Arsenal grounds. The library was "operated in conjunction with the Hartfield Public Library and Bookmobile, it is, to our knowledge, the only library provided by any ordnance plant in the United States" (Establishment of Boomtown Arsenal, 1943, p. 9). In addition to the rotating collection of books, the Arsenal also subscribed to several magazines and newspapers (p. 9). In the first year of operation, the library loaned 7,692 items to over 8,000 people (p. 9). The library represents a literacy institution made available by the government to workers and families that were housed inside the Arsenal's grounds.

The federal government provided funding to build or expand libraries in Fieldview and neighboring Hartfield. Within Fieldview, the government built a library inside the Community Building. Two of the people interviewed worked there at some point during the period of study, and seven (or 23%) spoke of frequenting the community library on a regular basis for school or leisure. I discuss more information that participants shared about the library later in this chapter as I discuss intersections across spheres of school and community.

School

Fieldview's School District was also provided with funding to expand to accommodate the influx of workers. As the Army designed the community in which it would operate and house workers (Elm), it included in the design an elementary school (K.T. Elementary School). The elementary school was located within the federal housing project, as was the community building, which housed the library. Several of the interview participants attended classes in this school building.

As with the library inside the Arsenal's grounds, the construction of the school and community library also represented making literacy-related institutions available for members of the community to encourage and facilitate literacy practices generally. Again, Grabill (2001) found that "literacy tends to be constructed in relation to the mandates of funding and policy interests (largely from government and industry) and to the goals articulated in large part by those interests" (p. 626). As Brandt, (2001) also observes, "literacy takes its shape from the interests of its sponsors" (p. 20). The inclusion of libraries and an additional school building articulated the government's valuing of print-linguistic literacies and the recognition that literacy was important to the Arsenal workers and their families.

Prufer (1982) indicates that a feature of Fieldview's school district's pedagogy was that it had regularly featured progressive education approaches espoused by Dewey (1938) and others; and it was noted as being one of the best school districts in the county. Progressive education espoused pedagogy that considered a student's own abilities, interests, and cultural identity and enabled students to learn based on their own living experiences in their community. As children of migrants entered the school district, they were exposed to this approach and benefited from it.

LITERACY PRACTICES IN SCHOOL

Thirteen (72.2%) of the people I interviewed were enrolled in the Fieldview School District for their entire elementary and secondary schooling, and fourteen (approximately 78%) earned their diploma from Fieldview. Of the 13 who attended Fieldview's school district for their entire experience, all but one reported practicing all of the listed forms of literacy practices: reading stories, books, and plays, and writing short papers, book reports, and poetry.

The interview question pertaining to reading and writing practices did not specify in what class or classes the activity occurred or was required. However, many referred to activities generally expected in English classes. Many spoke about a variety of reading practices and related activities. This diversity is represented by one's recollection of book reports and poetry memorization:

> H: Well, there again, we had textbooks that we read in that was sort of things that students do. We always, in English. In junior high school and high school English was a separate subject, and we did extra reading with written and oral book reports. We memorized poetry.

Some talked about how teachers encouraged them to read; two who spoke about their school activities in the 1950s called attention to the competition to earn reading certificates. Responding to a question about her use of the community library, one states,

> Mary: Yeah. The library had . . . I'd say, probably as many books and a school library, and probably had as many books as any other library and we were very much encouraged, and in the classroom, then, we would have the groupings of books that we were encouraged to read for nothing else but the certificate. . . . There was nothing more that we liked than the challenge of beating out anyone else in the class. I mean, if they were gonna get a certificate, by golly, you were gonna get one too. So that was our big encouragement.

> Me: So it was highly competitive?

> Mary: Our group was highly competitive and . . . that was it was highly competitive but yet we weren't mean about it. And the kids that . . . if you didn't do it he didn't do it and nobody razzed you about it. But there was a lot to keep up with. An internal kind of thing.

Another, responding to a general question about what reading or writing he recalled from school responds,

> Me: Do you remember any of the reading or writing activities from high school, in particular the assignments that stand out?

> Steve: The only things that stand out was the spelling. We would reinforce spelling. Spelling bees were reinforced. That there is quite a competition. No one most. I was one of those guys who play to compete. The notion of something in and speak something. Writing really didn't kick in for me personally until probably 7–8 grade, in that area.

This indicates that some students used the reading certificate program to motivate them to read more books over a given period of time. Such sponsorship espouses print-linguistic literacies.

Another person, though, recalled a map of Ohio that she had to draw as part of her school experience. In the following exchange, Claire clarifies that not all of her reading involved reading creative works:

> Me: These books were much creative pieces, then—poetry or stories.

> Claire: Yes, well, somewhat. But then there would be about a book about Ohio, and we would draw the shape of Ohio and talk about the different products and resources and that kind of thing.

Also, while none acknowledged specific recollections of them, picture books likely were part of elementary education experiences. So there were some visual literacies associated with education. However, the overwhelming majority of statements in interviews identified print-linguistic forms of literacy practices.

One person was a 2nd- and 3rd-grade teacher in the Fieldview School District during the period of this study, and she had an interesting comment about her observations of the children of migrant families. She explained that these children seemed to have considerably more difficulty with the reading activities and exercises. Her observations are shared in this exchange:

Me: Any other kinds of reading or writing training that you delivered, that you gave or any other kinds of activities that you can remember related to school or even outside of school?

GE: Of course we had workbooks. The children had to fill in their work-books; and it required help because some of them, especially the migrant workers who would come through. Those poor children didn't understand anything about what is a workbook.

Me: That's interesting.

GE: And different times they would start out with a workbook, and it gets so messed up that we had to take it away and give them another one to help them.

Me: That's interesting.

GE: But I think that there weren't too many migrant workers, and there were a few that spoke Spanish. I did not; and, so, they removed them from my room to the other third-grade teacher who did understand to teach them.

Me: Nonnative speakers? Do you remember any other kinds of difficulties that the children who were associated with the . . . with those who migrated to the area to work at the arsenal? Do you remember any other kinds of difficulties they had or any exceptional work that they performed in school again? Strengths and weaknesses?

GE: There are weaknesses in children who always lived in Windham, and there are weaknesses in the ones who migrated. I personally saw a lot more of the weaknesses in the ones that moved in, and I don't know if I was being biased or not, but . . .

Sponsorship Implications

In each interview, as people discussed various school-related activities, it is evident that they recalled most clearly the many opportunities they had to practice print-linguistic literacies. This illustrates another area in which print-linguistic literacies were sponsored more favorably over other forms of literacies, much as it was at the Arsenal.

LITERACY PRACTICES AT HOME

Print-linguistic literacy practices were emphasized and encouraged at home. Children and parents read newspapers, books, and wrote letters to friends and relatives on a regular basis. Fourteen people (almost 80%) reported some kind of writing practice at home, with most (50%) reporting several different types—diaries, journals, letters. Of the 11 who reported on their parents or siblings' writing at home, 9 reported that there was no writing practiced, however, it is possible that others' writing and reading were not visible to the interviewees. Ten people reported that they read a variety of materials—magazines, books, newspapers—and four (about 20%) reported that they did not read at all at home. Additionally, fifteen (84%) reported reading to others at home.

Seven people reported being encouraged in their literacy activities at home. Generally, either a teacher or a parent had the most influence on these activities. Such sentiments are reported in the following exchanges:

> Me: And you said that was during the 1950s. Any other information that you've recalled as we've talked about the home; reading or writing activities; or school, or any kinds of reading or writing instruction that you received?
>
> Jane: No. I just remember that we were expected to read a newspaper every day.
>
> Me: Your parents encouraged that?
>
> Jane: Yes. That was daddy's big thing; just to keep up on current events.
>
> Me: And would you talk about that; talk about what kinds of stories that were in the newspaper, then?
>
> Jane: It was . . . there were . . .
>
> Me: Any quizzes?
>
> Jane: Well, no . . . You knew you were supposed to do it and you did it. You respected your parents. You didn't question your parents back in my day.

And

> Me: And you said you struggled with reading. Did you read newspapers for . . .
>
> Beth: Yes, my mom made us.
>
> Me: Oh, she made you. What does that mean? Would she quiz you?
>
> Beth: No, she would discuss the article with us to make sure that we read it. She was very clever. And that's how I got to interact. I was . . . I read it too; and my brother was big into sports and he would read those sports, and my mom would ask him what he thought about this or that and I would be in there giving my two cents worth.

Parents encouraged children to read, learning print-linguistic literacies, by engaging them in quizzes or discussion to test reading comprehension.

Parents also encouraged print-linguistic practices by reading to their children, a practice Susan recalled:

> Susan: OK, there are . . . From those I can remember from elementary school reading the Dick and Jane stories. I can't remember literature or anything like that through junior high. I remember, for some reason, Ohio history, and I can kinda visualize it, the book. And all the things that we learned about Ohio, and it was very interesting to me. I learned about early governmental things that . . . how the Ohio began in the government way. My mother always read to us a lot at home. I think that's where we, my brothers and I, developed a love for reading; because she read a lot at home and I don't remember . . . mostly I think it was Bible stories. I can't remember a lot more than that.

Beth also spoke of receiving encouragement from her parents to practice writing skills.

> Me: Okay, did you maintain a diary?

> Beth: Yes, as I got older I did. And any time my parents took us on a trip, I would keep a log of what we were doing; and that, again, was my mom encouraging me to keep writing and doing some activities.

As I mentioned in a previous chapter, there was a security policy in place at the Arsenal that forbade employees from talking about their work outside of the workplace. In a few of the interviews, people acknowledged that their parents who worked at the Arsenal did not talk much about their work there. I alluded in that chapter to Steve's observation that neither he nor his father talked about work at home because of the policy. He said, "No, no. Pretty much. Pretty much in those days we had security clearances and the protocol in those days was to sort of leave work at work. Don't come home and talk about shipments or what you were doing."

Additionally, Susan acknowledged that her father never talked about his work when he was working at the Arsenal but did talk a little bit about it later in his life:

> Me: Did he ever talk at all about the kind of work that he did at home? Or did he rarely bring work home with him? Did he ever talk about kinds of training there? Was there or the kind of reading or writing he had on the job?

> Susan: At that time, no. He didn't bring any of his work home, and I'm sure I was like in bed or whatever when he would come home. He chose, for whatever reason, to work the afternoon shift in any position that he had. But, no, he didn't; not that I'm aware of it. Didn't talk to me.

> Although he had a tendency to talk about his work I do recall as I
> got older; but when he was at the Arsenal, no, I don't remember him
> talking about it at all.

Sponsorship Implications

Parents and teachers encourage children to read and write toward developing
print-linguistic skills. Parents and teachers recognize that such skills are important, and they try to encourage their children to develop such skills through
rewards and discussions. Consequently, parents and teachers, as Brandt (2001)
notes, act as sponsors of literacy, also favoring print-linguistic forms of literacy.
These practices reinforced esteemed practices at the Arsenal as well.

Literacy Practices in the Community

A variety of practices occurred in the community ranging from reading and
writing newsletters for civic organizations to writing instructional manuals
within 4-H activities, though none recalled community-sponsored adult literacy
programs. However, there were some community activities in which volunteers
practiced reading and writing. Three people reported participating in community programs—one was library sponsored and two were sponsored by civic
organizations. Generally, these involved reading or composing newsletters. One
example of this is from Steve, who participated in church-related newsletter
editing. When asked if he recalled any writing assignments from school, he
acknowledged that he could not recall any, but referred to writing activities that
were part of his church service:

> Those were those years I was doing things for the Methodist Church.
> Attended the Methodist Church in Fieldview; and some of us . . . we
> considered ourselves, the kids, trespassers and the congregation were
> debtors. So we went back and forth there doing the little newspaper there
> with the Methodist fellowship that I edited for a couple of years; and, of
> course, we brought those skills to bear to the Methodist Church.

While there was no organized community literacy program outside of school,
the civic and religious institutions supported literacy development. Further, as
Helen suggests, the school sponsored some of the literacy practices shared with
the church in that the children who produced the church newspaper brought
their school-obtained literacies to the newspaper.

Another participant spoke of her experience in 4-H and the writing activities
there:

> Helen: I don't remember other than the 4-H activities. I was in 4-H all this
> time.

Me: And did that involve any reading or writing?

Helen: Oh yes. Oh yes. We had our project booklets that we had to write in and read extra things about.

Me: And then you read about how to do certain kinds of things?

Helen: Oh, absolutely. Reading was just a part of my life.

Lucy also spoke of writing that she did connected with her participation in 4-H: "Well, I remember I was in 4-H, and I was the news reporter. So I had to write, you know, the minutes or whatever, you know whatever happened to be published in the paper. I did do that."

Half of the people interviewed (9) reported on experiences they had at the community library, though two of these were specifically school-related experiences. Five people reported leisure-reading activities pertaining to the library. Helen recalled the beginnings of the library in Fieldview and recalled frequently visiting it:

> But I might mention here that we didn't have a library in Windham until after, until after the housing project was in place. The community might have had a library in one of the old buildings near the old center of Windham, but that's not a very clear memory for me. I definitely remember going to the rural library and community building, and that was quite frequently.

These experiences are represented in the response Mary gave when asked about reading activities outside of school:

> On a personal level, we had no TV growing up, and it was an economic issue more than a stand that my parents took . . . more than anything. I think I was 12 when we got our first TV. Well, I can remember complaining terribly because the kids would talk about watching this show or that, but for me it probably made a huge difference because I read . . . all I had to do was to read. You know. We were friends with Mrs. Alexander who was the librarian, and the library was in the exact same place it is now only much smaller. In the community building. And, um, my mother didn't drive and my father worked afternoon shift, but Mrs. Alexander . . . I would walk across the field, probably first and second grade it started, and one day a week and I would spend from whatever time school was out—3 o'clock 3:30 whatever—until the library closed at 5 o'clock and just read by my choice and then Mrs. Alexander would take me home. And that was something that was an arrangement made between herself and my parents, but the thing that scaled the deal for her, I think, was that she stayed for dinner. My mother was a great cook, and she would have dinner there and my mother always fussed and have dessert, and I think that Mrs. Alexander liked that. So it worked out very well. I'd bring home a huge pile of books when I'd leave the library, and those books would be done by the next

week. So I personally read a lot because I had . . . in the winter . . . in the summer you're playing and outdoor things but in the winter there was nothing to do but, but to read.

The school district also facilitated visits to the library during school as part of a reading program. Teachers would take their classes to the library so the children could listen to a story read by a librarian. Betty spoke of her experience as a story-reader for this program:

> Me: Do you recall any library literacy programs—reading or writing pro-grams that the library held on a regular basis?

> Betty: Oh, yes. We always had the little children. We would have a reading program during school time. When the teachers would bring their class over to hear a story; so I was the story lady.

> Me: OK, good.

> Betty: And they would come maybe once every two weeks each class. But I would have a class every week that would come to the library, and I would read stories to them, and then we would talk about the stories. And they liked that.

While this experience seems one affected by economic dynamics for the family, it indicates a personal interest in reading for pleasure. It also recognizes a degree of sponsorship from the librarian and family friend, Mrs. Alexander. While the library acts as a literacy institution, the librarian appears to act as one who encourages attendance at the library and shows a personal interest in the students' literacy development.

The findings of library access are important to note. The presence of a library within the Arsenal grounds for employees and those officers and their families who lived on the grounds as well as the Hartfield library in a nearby community suggests recognition of the value of print-linguistic literacies.

Sponsorship Implications

As with experiences at school and home, participants recalled many oppor-tunities to practice print-linguistic forms of literacy in their community-related activities. People were even bringing school-related literacy experiences to bear on civic-related practices. As people are exposed to experiences that value print-linguistic forms of literacy in the various spheres of work, home, school, and community, they cannot help but perceive that such skills are valued generally and are necessary to advance in society.

INTERSECTIONS

It is important to consider intersections between the workplace and the spheres of home, school, and community, especially when studying any influence or sponsorship exerted by the Arsenal on practices in those other spheres. Print-linguistic literacies were esteemed both in the workplace as well as in the school, home, and community. While visual, aural, and experiential literacies were facilitated at the Arsenal, and visual practices were part of the school experience, an overwhelming amount of data show an emphasis on print-linguistic skills. This suggests that the government, especially, and Arsenal operators tried to sponsor print-linguistic practices over others within the local literacy ecology.

At the Arsenal, a variety of literacy practices occurred, and all were facili tated toward accomplishing a common geopolitical goal—national defense and victory in war toward peace and the maintenance of democracy. Print-linguistic, visual, aural, and experiential literacies were practiced at the Arsenal. However, the literacy hierarchy there favored print-linguistic skills. Also, at school, children were encouraged to learn print-linguistic literacies. Adults, even those who worked at the Arsenal, explicitly encouraged children to develop and practice print-linguistic literacies at home.

Adults in the community—parents and teachers—seemed to understand that print-linguistic literacies were valued over other literacies, even though all were practiced. Children were encouraged to develop print-linguistic literacy skills. While all modes were used to accommodate varied literacies in the workplace, people seemed to recognize an ideology that favored print-linguistic literacies, which is evident in a temporal study of the workplace archived materials and in the instruction students received at school. These suggest sponsorship of print-linguistic literacies over others by the government within the local literacy ecology.

As such, adults encouraged their children to develop those print-linguistic skills that they understood would enable the children to advance socioeconomically. Parents engaged their children at home in letter writing to relatives and friends. They also expected children to read books and newspapers to help the children do well in school settings. The government also supported this education-related literacy sponsorship through construction of the school and libraries.

There is an evident hierarchy represented in the modes of representation and related literacies in materials at the Arsenal. A variety of literacy practices occurred, yet the separation of modes emphasized across positions reflects a definite hierarchy that limited and regulated advancement. One cannot rise to certain levels without having a certain level of print-linguistic literacy. This esteeming of print-linguistic literacies carries into the community.

LITERACY SPONSORSHIP THROUGH TECHNICAL
COMMUNICATION PROGRAMS AND PRACTICES

Brandt's (2001) conception characterizes literacy sponsorship as "any agents, local or distant, concrete or abstract, who enable, support, teach, and model, as well as recruit, regulate, suppress, or withhold, literacy—and gain advantage by it in some way" (p. 19). I characterize the sponsorship associated with the technical communication practices at the Arsenal according to this definition. I address each attribute based on the findings reported in Chapters 5–7.

"Any Agents, Local or Distant, Concrete or Abstract . . . "

While parents, teachers, and librarians clearly acted in some capacity as sponsoring agents within the community, in the scope of this study, the government and contractors who operated the Arsenal acted as the primary local agents in that they were in close proximity and contact on a daily basis with those who worked at the Arsenal. Also, because of their proximity and considerable physical presence, the operators were a concrete agent of sponsorship for workers and the community.

The federal government acted as both a local and distant agent in that officials in Washington planned and directed certain operations while federal officials—military officers—held administrative positions at the Arsenal. It also was a concrete agent in that the officers at the site acted as representatives of the abstract federal government.

The two different operators of the Arsenal—APCO and Vulcan Tire—were also local sponsors. However, while APCO's sponsorship was difficult to separate from the government's influence, Vulcan Tire's sponsorship, clearly, differed from that of the government or Atlas' in that print-linguistic modes of presentation were favored over visual modes. However, this can be due to the number of employees who stayed on to work at the Arsenal and no longer needed the kind of training workers received in the 1940s.

" . . . Who Enable, Support, Teach, and Model"

The federal government and APCO embraced and facilitated the practice of multiple literacies, especially during WWII. The government, with help from industry leaders, developed the Training Within Industry program, which emphasized visual, aural, and experiential literacies, to accommodate low literacy levels among workers. This accommodation concerning anticipated literacy backgrounds of employees and the need for efficient retraining enabled increased production of war products needed to support the war effort. A certain amount of education was implied within each level of work at the Arsenal, but the ways training was provided and the ways in which training documents provided information speak volumes about differences in practices during WWII—before the accidental explosion and after it—and after WWII.

The modes used for training at the Boomtown Arsenal were affected by the following: labor pool and migration patterns, the literacy background of employees, the wartime economic environment that shifted labor skills from farming to war industry, and efficiencies needed to quickly prepare workers to do their work. The government espoused, and facilitated with the TWI program, training that minimized potential literacy differences among workers and created efficiency in training by emphasizing oral, visual instruction that integrated demonstrations and opportunities to practice the skills shown. Because employees came from varied backgrounds, Arsenal operators understood that there would be a need to accommodate related literate differences.

Mayer (2001) identifies two metaphors of multimedia learning: information acquisition, which facilitates adding information to one's existing knowledge; and knowledge construction, which applies to model building to facilitate cognitive guidance (p. 14). Training at the Arsenal focused on specific task-oriented information acquisition, disregarding the knowledge-construction-modeling metaphor. The emphasis was on training for a specific task as quickly as possible. Also, fewer people were needed during the Korean War, since operations there were on a much smaller scale. Consequently, the Arsenal operator managed to hire many of the people who worked at the Arsenal during WWII and stayed in the area, limiting the need for additional training. Such an approach values oral, visual, and experiential literacies over print-linguistic literacies.

Further, the manuals and SOPs used in training and as reference resources show changes in sponsorship attributes across periods. Steps presented in the G.P. bomb SOP (e.g., Figure 4.9 in Chapter 4), which was published in the 1940s, provide visual information with the print-linguistic text information in the spread. Also, the print-linguistic information is formatted to be visually accessible quickly. However, in all of the SOPs published during the 1950s, all of the print-linguistic text was placed before any visuals, though each visual can be identified with corresponding print-linguistic text. This difference suggests that the multimodal representations espoused by Mayer were recognized and esteemed during WWII, while they were devalued in the 1950s.

This study has found that the government and operators of the Arsenal esteemed print-linguistic literacies over other literacies. This emphasis is evidenced in the literacy hierarchy of the Arsenal, wherein practices at higher levels required people to have proficiency with print-linguistic literacies, while practices at lower levels required less print-linguistic proficiency and integrated more visual literacies. Further, the modification of print-linguist materials progressively toward lower grade-level literacy requirements suggests an effort to encourage print-linguistic skills as well as an understanding of the information.

" . . . as Well as Recruit, Regulate, Suppress, or Withhold Literacy"

An overemphasis on the visual and experiential can create a precarious situation. While manuals were distributed, these were rarely used other than as reference guides in case someone forgot a specific task. As evidenced with the bomb explosion incident, when a safe protocol is forgotten or set aside because of time issues or product-quality issues, it compromises any efficiencies and places workers at great risk. Workers need to be encouraged to read any written materials that may reinforce safety guidelines associated with operations. Further, such materials could have provided information that was not clearly articulated orally to help workers understand what specific precautions to take. Manuals published after the explosion accident did this.

Supervisors and administrators, those in higher positions, were expected to have print-linguistic literacies that included higher grade-level expectations than workers likely had, in addition to visual literacies. One could not advance within the organization without these skills. In this sense, the literacy hierarchy at the Arsenal represents the literacy-related ideology that privileged print-linguistic modes over visual modes of representation.

Print-linguistic literacy practices were also actively encouraged throughout the community. Data associated with the library usage, school instruction, and practices at home and in community organizations emphasized print-linguistic skills. People saw the value of these literacies in possibly advancing themselves and their children. These practices favored print-linguistic literacies, even though a program that facilitated multiple literacies was practiced at arsenals around the country during WWII. Subsequent to WWII, however, the print-linguistic ideology seems to have reemerged as the favored approach as materials at the Arsenal emphasized those literacies over the multiple literacies practiced during WWII.

" . . . and Gain Advantage by It in Some Way"

Brandt (2000) observes that the literacy environment of pre-WWII brought about the literacy crisis of WWII. She characterizes the 1930s literacy ecology as "regional stratifications" of the literacy economy brought about by the geographic and economic make-up of the United States—largely a farming economy that affected how children were taught and how adults valued literacy practices toward farming applications. Fieldview was a farming community prior to WWII. The training associated with the military-industrial economy of WWII provided instruction in various technologies that would continue beyond the war period and become part of the national and, eventually, global economy. This connection illustrates a link between literacy and the social, political, and economic needs of the country at a given time period.

Finally, generally, it appears that the literacy practices of the workplace did not negatively affect literacy practices within the community—at home, at school, or in civic organizations or in leisure-reading activities. Children were encouraged to learn print-linguistic literacies even as they practiced visual literacies such as drawing state maps. However, the practices clearly suggest an understanding that print-linguistic literacies were esteemed over visual or other literacies. Brandt (2001) also notes that print-linguistics literacies were valued during much of the 20th century.

Much as a certain level of education was required to advance within the Arsenal's literacy-related hierarchy, so too did that literacy-professional advancement relationship extend beyond the Arsenal.

Groups That Benefited

While the nation benefited from the multiliteracy practices at the Arsenal in terms of ammunition supplies available for the war effort, clearly, people with low print-linguistic literacy were able to acquire a job at the Arsenal; this provided them a livelihood. Further, just as Brandt (2001) observes that print-linguistic literacies were required to advance economically throughout the 20th century, the Arsenal encouraged print-linguistic literacies by using them to facilitate professional advancement there as well as helping children in the community to acquire such skills to position them to be able to succeed economically as well.

Groups Negatively Affected

While it appears that most groups benefited from the government and contractor's sponsorship, the literacy hierarchy I observed limited movement across levels of the organization. Those with low literacy levels could not advance to higher positions in the Arsenal's management structure. Also, as described previously, employees who handled explosives were placed in a dangerous position because they had limited information because of the separation of modes of representation and expectations of literacy therein.

CONCLUSION

The study detailed in the chapters so far provides analysis illustrating how technical communication practices are sponsored and sponsor literacy broadly. The next chapter describes more recent applications of TWI in technical communication practices and how its sponsorship continues to affect technical communication practice and pedagogy as well as literacy practices outside of the workplace.

CHAPTER 8

Current Applications of Training Within Industry: Continued Sponsorship of Technical Communication

As described in previous chapters, the Training Within Industry (TWI) model integrates attributes of multimodal forms of representation in multiple ways, all of which are encouraged in technical communication pedagogy and practices. While it includes print-linguistic text in documentation associated with instructions and proposals, it encourages graphic representations of this information in these documents to illustrate comparisons and categories related to them and to provide a visual representation of the processes and behaviors. It also integrates experiential practices through hands-on learning.

The same approaches to hands-on training are still used today. However, new media technologies also facilitate applications beyond print-linguistic documents such as manuals and SOPs and real-time training on-site. They also include online training, video instruction, and development of websites that provide information. This new media also affects the attributes in TWI that I discussed in Chapters 4 and 5.

TWI is increasingly espoused in many companies today, including Toyota, IBM, Micron Technologies, and New Balance, among others (Liker & Meier, 2007; TWI Learning Partnership, 2009). Indeed, the TWI institute lists companies across several different industries currently using TWI; these industries include information technology, consulting, energy, aerospace, automotive, manufacturing, and health care (TWI Institute, 2012). These applications range from using modified versions of the original forms to the use of web-based training programs.

This chapter describes a few of these applications, emphasizing the multimodal representations in documents, artifacts used regularly, and in training and applying to this discussion the concept of literacy sponsorship associated with technical communication. Multimodality is used in many forms in technical materials, and the government sponsors such literacies through its sponsorship of TWI within the Manufacturing Extension Partnership (MEP) program.

The TWI-related materials and processes of training and improvement integrate all of the forms of representation identified by the New London Group (1996) and principles of multimodality presented by Mayer (2001), as mentioned in previous chapters. Specifically, Job Instruction, targeting supervisors and workers, integrates visual, aural, behavioral, and spatial modes for workers, while it includes all of these as well as print-linguistic modes that integrate visual attributes (tables) within supervisor training. Job Methods integrates mostly print-linguistic and visual modes within the use of tables and diagrams to represent information. Job Relations and Job Safety programs include some elements of print-linguistic modes in the process of collecting information, but emphasize oral/aural modes in talking with and listening to employees. The interaction between these modes is discussed in this chapter relative to the current applications, emphasizing Job Methods and Job Instruction.

I present a variety of examples of TWI in its current form in this chapter as well as how its attributes are used in forms of training/education not explicitly linked to TWI. While not as detailed an analysis as the case study presented in the previous chapter, it shows how companies are using TWI's multimodal rhetoric in various applications today through government sponsorship. It also demonstrates the proliferation of the use of multimodal rhetoric associated with TWI in workplaces beyond explicit application of TWI. Some of the differences that contributed to the bomb accident are minimized as multimodal forms of communication occur over various levels of organizations.

THE MANUFACTURING EXTENSION PARTNERSHIP AND TWI

Dinero (2005) notes that after World War II, TWI's philosophies were no longer esteemed and practiced in the United States since there was little need for efficiencies connected to it. However, he also points out that TWI flourished overseas in many countries, including Japan. Indeed, Toyota used it prominently in its production systems. However, as demand for efficient operations returned to the United States, so did TWI as an esteemed model. The federal government supported this movement.

The Manufacturing Extension Partnership (MEP) was initiated in 1989 by the federal government in an effort to encourage businesses to implement efficient programs. Scholarship in the field of management concerning continuous improvement systems and lean manufacturing emerged in the 1990s, and MEP was connected with those philosophies. Since the mid-1990s, the program has grown to include approximately 350 centers (Schacht, 2009, p. 2). Each state in the United States has at least one headquarter (or Center) related to the MEP program. These all integrate programs explicitly linked to lean and continuous improvement, including Six Sigma, a program aimed toward reducing quality control errors. Most of the affiliated offices integrate TWI, however, not all do.

As I mentioned in Chapter 1, the MEP program, and specifically TWI, are used in a number of industries today. However, many of the philosophies related to TWI also exist in business applications not explicitly linked with MEP or TWI.

Further, many of the Centers in each state have direct and explicit connections to institutions of higher education in those states. For example, Purdue University is Indiana's MEP Center, and North Carolina State University is North Carolina's MEP Center. Other institutions, while not as closely affiliated so as to have the state's Center, provide training programs in conjunction with a state-certified Center; Texas Tech is one such institution. So companies are sending their employees to MEP programs offered as outreach efforts by institutions. Coincidentally, many of these institutions also have technical communication programs. This connection can easily include internships between students in technical communication or professional writing programs in which students practice the multimodal rhetoric of TWI. Such internships integrate elements of literacy sponsorship as well.

Manufacturing

Similar forms of TWI are used today to show problems with processes and possible solutions. The diagramming used within TWI's Job Methods program help to illustrate specific steps, and a reader can visualize problems and solutions when comparing two diagrams. The ABB-SSAC company used TWI explicitly to develop a more efficient method for a particular process. Figures 8.1–8.3 show diagrams of a process that involved the use of three different buildings, and the reader can visualize through these diagrams how confusing and inefficient the operation was with the process crossing in different directions at various points.

Figure 8.4 shows the newly designed process, which uses a single building and appears to be much more efficient, using a single building and with the flow of work moving in a single direction.

The diagrams are specific applications of the Jobs Methods approach of TWI— diagramming a process to find efficiencies—however, it also shows the application of multimodal representations in materials that would have executives as the primary audience. At the Arsenal, documents read by executives tended to use few graphics; in this case, graphics are used more often to contribute to an executive decision process.

Print-linguistic text is integrated as labels along with the visual diagram showing the entire processes and the problems with it and improvements to it. The information associated with the breakdown sheets and diagrams is added to proposals and enhances the proposal by showing the reader specific step-related drawbacks associated with the current process and benefits of the proposed process.

Technical communication scholarship addresses use of diagrams for the purpose of identifying problems, and such applications are illustrated in technical

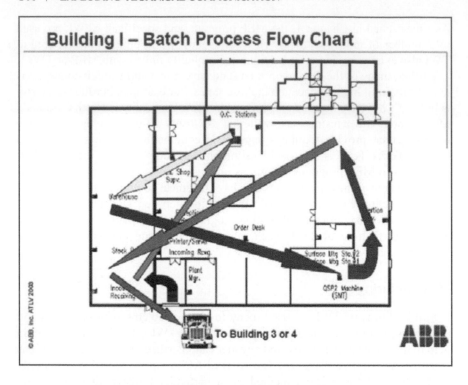

Figure 8.1. Diagram of original process—Building 1.
Source: ABB (2008). SSAC operation-Lean Journey. TWI Summit
http://www.twisummit.com/2008presentations/ABB%20-%20TWI%20
Orlando%20may%202008.pdf accessed April 12, 2010.
Used with permission from ABB.

communication textbooks (Johnson-Sheehan, 2012; Markel, 2010). While TWI's initial applications pertained to manufacturing, they have expanded to include health care.

Healthcare Application

Miller (2005) describe the implementation of TWI at two healthcare facilities: Virginia Mason Medical Center of Seattle, Washington, and ThedaCare, based in northeast Wisconsin. Graupp and Purrier (2012) provide a detailed discussion of it at Virginia Mason Medical Center. I describe some of the particular attributes of the multimodal rhetoric at Virginia Mason Medical Center used to improve surgical procedures and nurse scheduling.

Multimodal forms of representation are evident in specific equipment designs as well as in documents and training. Staff at the Center redesigned the layout of

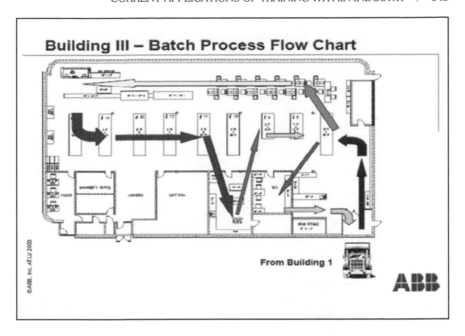

Figure 8.2. Diagram of original process—Building 3.
Source: ABB (2008). SSAC operation-Lean Journey. TWI Summit
http://www.twisummit.com/2008presentations/ABB%20-%20TWI%20
Orlando%20may%202008.pdf accessed April 12, 2010.
Used with permission from ABB.

a board used to hold surgical equipment. Figure 8.5 shows what an Anesthesia Shadow Board looked like before changes were made.

Equipment is scattered about the board in what appears to be somewhat of an organized manner. Certain equipment is set to the left, a particular device is placed in the center, and syringes, needles, and medications are positioned to the right. However, the staff was able to implement a better design to the board that included labeling items more clearly to improve efficiencies. Again, though, the use of visual and text are combined to help one understand where specific items are located quickly.

Figure 8.6 shows the modified board, which includes a template labeling where specific items are to be placed. Each item is placed in a clearly marked position, and staff can immediately locate particular tools as needed. Each section of the board includes labels, and the user comes to learn to expect certain equipment in certain locations, much like the reader of the bomb manufacturing SOP shown in Chapter 5 would come to expect to find procedural information

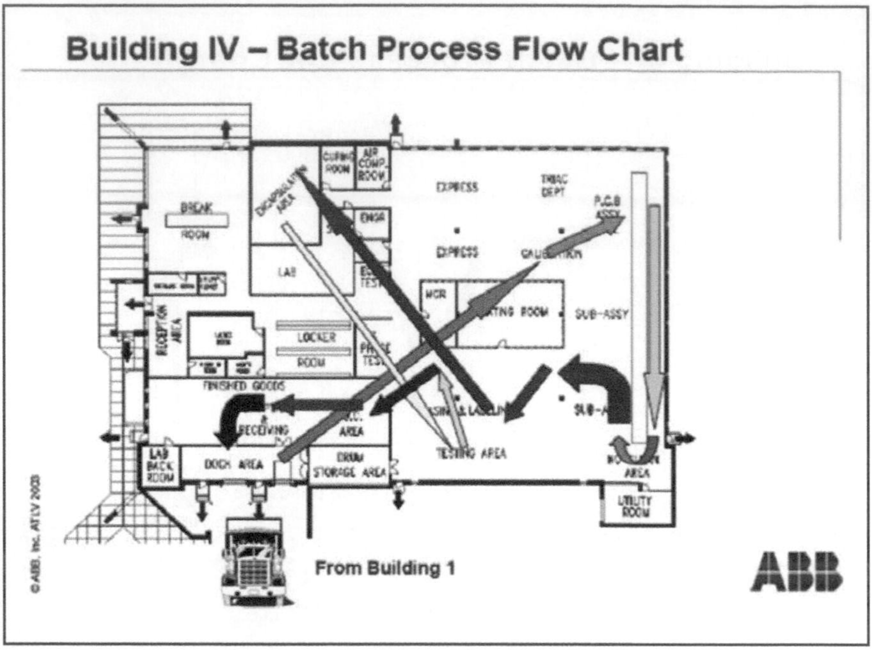

Figure 8.3. Diagram of original process—Building 4.
Source: ABB (2008). SSAC operation-Lean Journey. TWI Summit
http://www.twisummit.com/2008presentations/ABB%20-%20TWI%20
Orlando%20may%202008.pdf accessed April 12, 2010.
Used with permission from ABB.

in one particular location of each page and safety information in another part of the same page.

The combination of labels on the template and a visual organization of the board make it a more efficient setup than the previous board. Again, the combination of visual as well as print-linguistic text enhances the ability of the user to interact with the board. The template acts to guide users' understanding of the position of specific information, much as template forms functioned to guide organization of information for supervisors and inspectors at the Arsenal. Northey (1990) observes that writing at low levels of management tends to be template, while writing in upper management tends to be more customized. So templated forms such as these help minimize cognitive requirements of users, yet they also integrate visual representations that print-linguistic text could not capture without thick description.

Given the context in which the table is used—surgery—the template allows very quick access to each piece, so the design suits the particular rhetorical

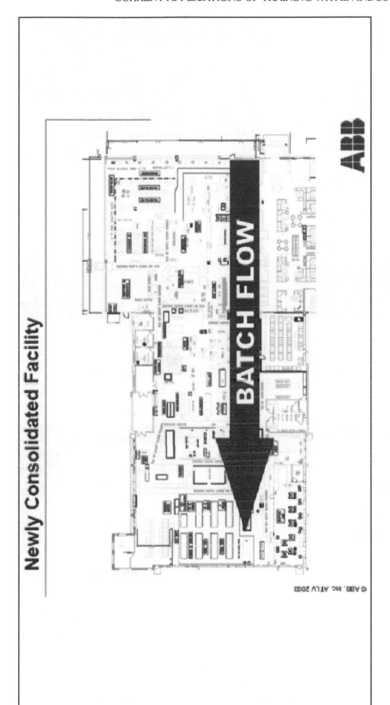

Figure 8.4. Diagram of new process.
Source: ABB (2008). SSAC operation-Lean Journey. TWI Summit
http://www.twisummit.com/2008presentations,ABB%20-%20TWI%20Orlando%20may%202008.pdf
accessed April 12, 2010. Used with permission from ABB.

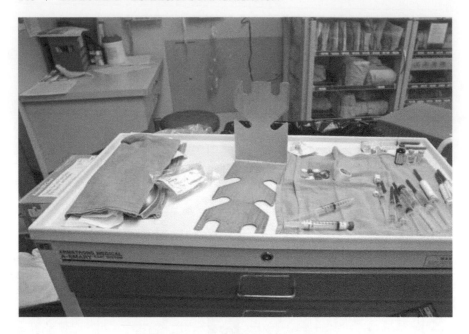

Figure 8.5. Previous shadow board.
© 2011 Virginia Mason Medical Center.
Images used with permission of Virginia Mason Medical Center.

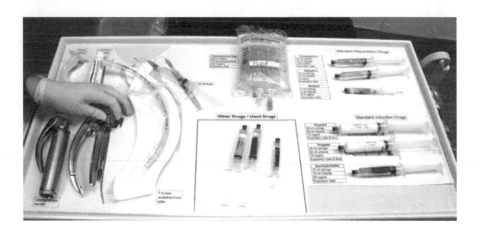

Figure 8.6. Templated shadow board.
© 2011 Virginia Mason Medical Center.
Images used with permission of Virginia Mason Medical Center.

attributes of purpose, audience, and context. Nurses will understand the terminology associated with each label, and the template allows for a prescribed protocol associated with the layout of the equipment. Once a nurse learns that layout, he or she can quickly locate a specific item and hand it to a surgeon or address an emergency him or herself. Such efficiency would be greatly advantageous in a life-threatening setting.

Another form shows the processes involved in assigning nurses to patient rooms. However, the diagram is placed with a table in which problems are listed with print-linguistic text, a combination of print and visual representations like those in Figure 8.7 listing the steps in a process toward improving the process. The diagrams showing the "before" and "after" visualization illustrate the problems and improvements, providing a visual representation for the reader.

This information would be viewed by the team that is trying to improve the process in conjunction with oral narration describing the dynamics represented in the diagrams. But print-linguistic skills are used more elaborately than in just labeling items, and they are of value in describing specific items while the table provides a means by which to categorize certain attributes of the steps or processes, highlighting problems. The diagram shows what these listed and described items look like in practice to facilitate better visualization of the entire system, working in conjunction with the oral narration.

Print-linguistic skills are valued but coupled with visual elements of tables in much of the TWI practices regardless of the level of worker or management. However, Virginia Mason Medical Center also ascertained an inefficiency in the print-linguistic information in a breakdown sheet and refined it toward more concise wording. Figure 8.8 shows a current steps table for hand hygiene from the Virginia Mason Medical Center.

Each step is listed with specific points related to it and reasons for the points. This table details not only the steps but particular attributes of each step. This allows the reader to better understand each step and its relationship to others in the process. However, this table also represents an inefficient use of wording in the breakdown of the task. Figure 8.9 shows the revised sheet that VMMC uses, only with more concise wording.

While there is less wording to describe the steps, one of the elements of TWI is the valuing of multimodal training; so one would also receive hands-on instruction showing them how to perform these steps while also receiving verbal information about the steps as well. Trainees would see the actual hand motions involved while hearing the instructions.

This information is for a supervisor/trainer audience and integrates visual as well as print-linguistic literacies. It considers audience attributes in that it uses few specialized terms, and those that are used would be understood by a medical practitioner. The form is used in training, so no consideration of speed is included; the context does not require it.

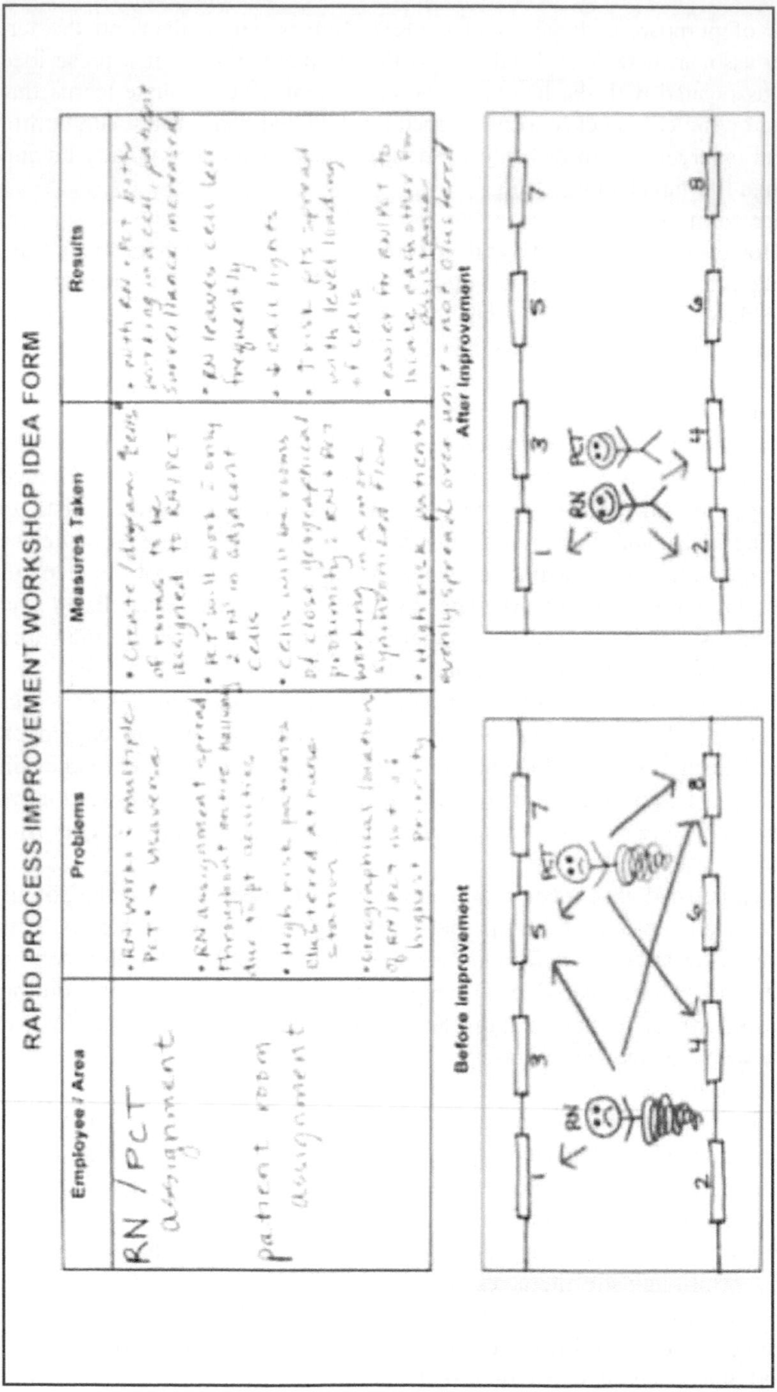

Figure 8.7. Diagram for process improvement for room assignments.
© 2011 Virginia Mason Medical Center. Images used with permission of Virginia Mason Medical Center.

MAJOR STEPS	KEYPOINTS	REASONS FOR KEYPOINTS
Step #1: Identify the need for clean hands	Remove artificial fingernails or extenders when in direct contact with pts or their	Artificial nails house germs that can be passed on when you touch pts
	Clean hands whether or not you use gloves (i.e. before putting on gloves & after removing gloves);	Gloves are not a substitute for cleaning hands because gloves don't completely prevent germ transmission
	Before direct contact with pt, pt's environment or equipment	Protect the pt against harmful germs carried on your hands
	After direct contact with pt, pt's environment or equipment	Protect yourself & the health-care environment from harmful germs
Step #2: Inspect your hands to determine best cleaning method	If not visibly soiled, use alcohol-based gel	Cleaning with gel is faster, more effective, and better tolerated by your hands
	Visibly soiled hands or hands with fecal contamination require washing with soap & water	Dirt, blood, feces or other body fluids are best removed with soap & water (C.diff spores are not killed with alcohol-based
Step #3: Use enough product to cover all hand surfaces & fingers	GEL: Cover all surfaces with a thumb nail-sized amount	Friction & skin contact are required to remove germs
	WASH: Wet hands with water, wash with enough soap to cover all hand/finger surfaces	
Step #4: Spend enough time cleaning your hands	GEL: vigorously rub until product dries on your hands	Antiseptic action is not complete until fully dried (approx 15
	WASH: a minimum of 15 sec (the length of singing "Happy Birthday to You")	At least 15 sec is needed to ensure complete coverage of hand surfaces
	Use paper towel to turn off water faucet	Prevent transfer of germs from faucet onto clean hands
Step #5: Let your hands completely dry	Moisturize hands PRN with lotion available through Central Supply	To minimize contact dermatitis without interfering with antimicrobial action
	Put on gloves after hands are dry	Skin irritation may occur if moist hands come in contact with glove material
Step #6: Perform task with clean hands	Task is done immediately after cleaning hands	You may be distracted & touch unclean surface with clean hands

Figure 8.8. Job instruction breakdown sheet for hand-washing.
© 2011 Virginia Mason Medical Center. Images used with permission of Virginia Mason Medical Center.

Version 1.0
07.07.09

JOB INSTRUCTION BREAKDOWN SHEET – HEALTHCARE

Task: __Hand Hygiene-Washing__

Supplies: Soap, Running Water, Disposable Towel

Instruments & Equipment: _____

IMPORTANT STEPS	KEY POINTS	REASONS
A logical segment of the operation when something happens to advance the work	Anything in a step that might – 1. Make or break the job 2. Injure the worker 3. Make the job easier to do, i.e., "knack," "trick," special timing, bit of special information.	Reasons for key points
1. Wet hands	1. Without soap	If soap first, it rinses away
2. Apply soap	1. Cover surfaces	
3. Rub hands	1. Palm to palm 2. Palm to backs	
4. Rub fingers	1. Thumbs 2. Interlocking 3. Backs of fingers to palm 4. Tips to palm	
5. Rinse	1. Leave water on	
6. Dry	1. Use towel to turn water off	Prevent recontamination of hand

Figure 8.9. Job Instruction Breakdown Sheet for hand-washing.
© 2011 Virginia Mason Medical Center.
Images used with permission of Virginia Mason Medical Center.

Further, the trainer supervises the trainee as the trainee performs the tasks. The supervisor shows the trainee how to do a given task, talking through the important attributes of the process. This provides visual and oral instruction showing even the spatial and behavioral dynamics involved in the process. The trainer then has the trainee do the same task, talking through the important steps, to ascertain that the trainee has learned it. This allows the trainee to experience the gestures and other behaviors associated with the task. The various modes used in the training reinforce each other to enhance learning.

As suggested with these applications, TI has reemerged in various fields toward increasing efficiencies in learning and improving work. However, TWI also is supported through federal programming. The next section details that effort.

Manufacturing Extension Partnership and TWI as Sponsor: "Success Stories"

The federal MEP website (www.MEP.gov) lists a number of other "success stories" of companies implementing TWI and the qualitative and quantitative benefits the company realized with that implementation. I provide a summary of some of those success stories here because they suggest that the rhetoric of TWI is effective, and they also suggest the impact that government sponsorship of such a program has on technical communication. It is one thing to describe rhetorical practices, as is often done in multimodal and technical communication scholarship, but it is another thing to present quantifiable information about the impact of that rhetoric and sponsorship therein. By providing quantitative information about the impact of a given form of technical communication suggesting its success an entity is encouraging others to use it, thereby sponsoring it.

Anchor Packaging, a food packaging company in Arkansas, acknowledges that by implementing TWI, it has "reduced training time by 30%, reduced turnover by 20% and improved quality by 15%." Electronic Systems, a South Dakota company that provides electronic manufacturing services to companies in medical and telecommunications industries, reported using TWI to reduce defects from 75% to 0%. Giant Snacks, a North Dakota sunflower seed company, reported reducing the learning curve for new employees by 66% with the TWI program. Medtronic Physio-Control is a Washington State-based manufacturer of defibrillators. The defect rate for one of its defibrillators was so high that the FDA suspended its U.S. sales of that defibrillator. In response, Medtronic implemented TWI's training program and the defect rate dropped dramatically, leading the FDA to lift the suspension. Finally, the Trane Company, a Colorado manufacturer of water coolers for air conditioners, noticed that each time it implemented a new continuous improvement program, defects would be down and productivity up for a short period of time, after which these gains began to decline. Upon implementing TWI programs, though, it noticed a sustained improvement in productivity of 40%.

Again, these TWI "successes" are reported by the MEP website along with several other success stories from other companies that have implemented TWI. The success stories on the site include testimonials about the other programs in the MEP program too, however I reviewed only those that used the TWI program. I list them here because they suggest that the multimodal rhetoric of TWI's training program generates favorable results important to business and industry. As such, the site itself also acts as a mechanism of sponsorship since it calls attention to the benefits of the program's multimodal technical communication practices.

APPLICATIONS BEYOND TWI

Many recent applications of training and education are facilitated by new media and engage the same multimodal principles of TWI and are encouraged in technical communication pedagogical scholarship. These include online tutorials for providing instructions and the use of simulators for training purposes. Online tutorials come in many forms, including straight video that includes narration, video with text, and text that includes narration. Simulators range from two-dimensional formats to three-dimensional environments such as *Second Life*. All use multiple forms of representation to make and enhance meaning and integrate the principles of TWI. Scholarship in professional and technical communication also has theorized the multimodality associated with video products (T. Johnson, 2008; Jones, 2007; Wysocki, 2001, 2003) and virtual worlds (deWinter & Vie, 2008; Remley, 2010a, 2010b, 2012; Vie, 2008a, 2008b).

Video Demonstrations

A popular form of online tutorial is the video demonstration. Generally, narration accompanies the video, which shows a viewer how to perform a particular task. Several examples of this are available on the site YouTube.

Alternatively, there may be some text that appears during the playing of the video. For example, some online banking services sites have such a video to show users how to make payments using their services. In the video, the interface is shown and steps are presented using text. As the video progresses, different parts of the payment screen interface are shown, and text appears as various mouse actions are illustrated. This kind of video demonstration integrates visual and print-linguistic text as well as some spatial elements, showing what the interface looks like while integrating instructional information. Viewers of any level can understand the information given the combination of visual and print-linguistic modes of representation.

Interactive Templates

Another form of online tutorial is the interactive template. Generally, it involves a print-linguistic interface that provides information about how to do a given task and what the interface would look like, then it invites a user to perform some similar task. The user then inputs information similarly to how the template described and sees what happens. An example of this is the template provided by the W3 site to show how to use Cascading Style Sheet (CSS) code to design web pages (www.w3.org). The textual information explains the various codes to enter and how to enter them. The user is then invited to enter different codes in the template using similar code formats. Another window in the template then shows what the page screen would look like resulting from the particular codes the user entered. Such templates integrate print-linguistic text, visual, as well as behavioral elements to help one understand how to perform a task. However, this is able to be replicated by placing the CSS code file next to the page file as one develops a page. Design tools such as Dreamweaver also facilitate a coding screen that appears in conjunction with the resulting page screen so that users can see what effect certain code will have on the web page.

Such tutorials represent not only the "learn-by-doing" approach of TWI, but also integrate elements of co-multimodal redundancy, reinforcing textual and visual information. CSS coding, for example, is formatted in a visual way so the viewer can quickly access certain sections of information. The page file, then, shows not only the spatial and visual attributes coded, but also reinforces the mental image the designer visualized as he or she inputted the coding.

Simulators

Simulators allow users to interact with a replication of the actual environment and context in which a particular activity would occur. As such, there are a number of multimodal attributes at work. Because the user sees things in multiple 3-dimensions, as one would normally see them, it facilitates visual and spatial modes. Because the user engages specific objects and tools, it engages tacit, behavioral attributes as well. Generally, an instructor is nearby to provide information orally, so there is an aural attribute as well. If the simulator pertains to a particular screen that includes print-linguistic labels, one also experiences print-linguistic text. I describe the multimodal rhetoric links with TWI's approaches for various kinds of simulators.

SECOND LIFE

A popular simulator for education in general and, increasingly, professional education, is *Second Life* (SL); a three-dimensional virtual environment in which users can interact with objects similar to a real-life experience. It has been compared to a video game, and while users can create game environments in it, game theory is not built into the default system such as systems like World of Warcraft. Stephanie Vie (2008b) observes that, "*Second Life* has grown in popularity among educators because of their interest in the possible pedagogical uses of video and computer games, including their potential ability to strengthen students' critical media literacies." SL has become a popular space for medical training. Mesko (2008), for example, describes a space in SL wherein medical students can engage in a realistic patient visit experience. Medical students enter a facility and ask to see a patient from a list of available patients. After the student's avatar washes his or her hands in-world, the student then is shown the patient's medical chart. The student has to respond to a few questions about the chart, making preliminary diagnoses at a nearby file cabinet based on the information on the chart (Figure 8.10).

If the student passes that test, the student is able to request particular tests to further understand the patient's condition.

Such a simulation also considers the audience, purpose, and context. It is a training setting with students who are learning how to behave like a doctor. It takes students through the steps associated with a patient visit without placing the student in a position in which someone's life is in danger, removing stress

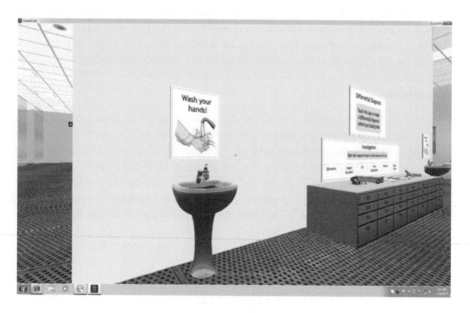

Figure 8.10. Imperial College *SL* facility.
Used with permission by Imperial College London.
Island objects designed by Maria Toro-Troconis,
E-Learning Strategy and Development Manager, Imperial College London.

from the learning environment. However, the student interacts with medical terminology as he or she normally would as he or she conducts the visit.

Mollman (2007) also describes an SL application used by Orkin to train workers to inspect homes for pests. An avatar enters a home and moves about it checking in particular places for evidence of pests. Hudson (2010) describes an application related to training students to perform border inspections within a justice studies program at a college near the U.S.-Canada border. He states that students using the SL simulation did 39% better on specific performance measures than students who did not use the SL simulation, suggesting that the simulation contributed to better learning of specific skills (p. 117).

In each case, the SL environment is used to simulate the actual environment one would experience in a particular role and enables the user to perform certain actions in a virtual environment. Using the virtual environment eliminates any real problems that could occur from accidents one may encounter as they learn the task. So it engages trainees much like the learn-by-doing approach of TWI while facilitating demonstrations too. Simulators like SL engage visual, aural, spatial, and behavioral modes of representation in an immersive experience similar to actually performing the task in the real world.

Further, demonstrations in SL can be recorded like videos and shown subsequently. This allows such demonstrations to be scripted, including gestures an avatar would integrate. As such, video pertaining to virtual environments (generally referred to as "machinima" video) apply attributes of dramatically scripted demonstrations.

OBJECT-SPECIFIC SIMULATORS

Other examples of simulators include police training simulators, mining simulators, and flight simulators (Figures 8.11–8.13), all of which allow one to interact within an environment that simulates one in which they would be working as they train—training by doing. Figure 8.11 is a simulator that helps students learn how to control a plane in various conditions, while Figures 8.12 and 8.13 pertain to another simulator that helps students learn landing approaches. With the approach simulator, students also receive a graphic printout of the path they took, which is represented on a flight map. With such simulators, one uses specific tools that one would use in real life, and the simulators are able to show how certain sequences would affect an outcome, desired or undesired, so the user can see what would happen without a catastrophe occurring. Students receive about 30 hours of training with the simulators, in addition to receiving instruction in the air with a pilot instructor, before being placed in a control position in a real airplane.

Yet another example of simulators is new simulation technology for training air traffic controllers (ATCs). Many ATC programs break down the responsibilities associated with air traffic control into different lessons, similar to the

Figure 8.11. Airplane simulator view. Photo by Dirk Remley.

Figure 8.12. Exterior of second airplane simulator. Photo by Dirk Remley.

Figure 8.13. Interface of second airplane cockpit simulator.
Photo by Dirk Remley.

TWI approach of breaking down a job into smaller units to facilitate quick training. Further, simulators of varying types engage students in hands-on learning. Figures 8.14 and 8.15 illustrate two different kinds of simulators related to air traffic control training: the actual landing and take-off operations.

This is an overhead view of an airport and its runways. Lights representing aircraft would appear on the screen; there are no planes on runways in this particular image. However, a visual of the airport's runways is coupled with this screen, as shown in Figure 8.15.

With the experience, students get two different visual representations of the runways and can monitor aircraft moving about the runways accordingly. The experience also includes considerations of the rhetorical dynamics of the situation: audience, purpose, and context. A student is not involved with a real aircraft, so no one's life is in danger as students interact with the realistic tools they would use professionally. Also, while these different procedures are part of any air traffic controller's responsibilities, students learn each process separately in different courses. This is similar to the way TWI broke tasks down into smaller tasks to facilitate learning a particular task more easily. The capstone course in one institution's air traffic control program involves control tower management in which all responsibilities are integrated.

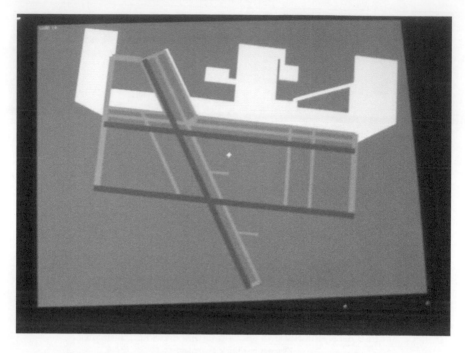

Figure 8.14. Overhead view of runways. Photo by Dirk Remley.

Figure 8.15. Combination of computer display and tower simulation view.
Photo by Dirk Remley.

In each simulation case, though, the principle of learning-by-doing applies, and this principle engages experiential learning as well as visual, aural, spatial, and behavioral modes of representation.

The TWI program continues to impact technical communication practices today through the sponsorship of the MEP and TWI Institute, which provides training of supervisors to facilitate TWI training of employees. Much as the government and private industry teamed to form the initial TWI program, the same combination joined to form the MEP program (Schacht, 2009).

The real, face-to-face, hands-on demonstrations and practice approaches to training are still a part of the TWI program, however new media technologies have encouraged similar approaches that facilitate asynchronous training and instruction. Video that integrates print-linguistic text and/or audio narration can be used to demonstrate tasks. Also, virtual environments that can replicate certain spaces can be used to help users experience and practice actual behaviors they will use on the job. Even materials that integrate print-linguistic text can become interactive to allow users to practice performing certain tasks themselves and see the results of their own actions.

The process-improvement materials used in TWI still encourage integration of graphics to show specific processes and inefficiencies as well as ways to improve it. New media enables employees to provide videos and graphics in presentations in which they propose new methods toward process improvement as well. Such tools integrate multimodal elements for audiences at any level of an organization, reducing the perception of a literate hierarchy.

Even outside of the TWI model, multimodality is embraced in industry and education. Much like the rapid growth of web-design companies that occurred in the 1990s and early 2000s, there is a general growth of companies that develop and create online training tools such as video and interactive tutorials, including machinima videos developed with 3-D virtual environments. For example, DemoWolf develops customized tutorials in Flash, video, and text/image (like step-by-step instructions that include a screen shot) formats for companies. Also, 3-D virtual platforms are being used for business and industrial uses, ranging from facilitating meetings, presentations, and training that include video training (Wagner, 2007). The existence of over 20 companies that produce machinima videos (Linden Labs, 2010) suggests demand for machinima video design skills.

The one machinima video that I have come across that is closest to duplicating the TWI instructional principles outside of the TWI clientele is one developed by a company that provides instructional materials for youth soccer leagues.

Soccer Video Product

The Challenger Sports Corporation produces instructional videos to help children and teens learn various soccer skills and advanced moves. As a youth soccer coach for my daughter's soccer team, I wanted to have my daughter watch a video instructing on a variety of soccer skills, and I came across their video at the local library. The video includes a series of short instructional machinima videos in which a virtual boy or girl demonstrates a given skill. Each segment is set up applying a number of elements of the TWI instructional approach.

Each segment begins with a male narrator introducing the particular skill that will be demonstrated by identifying its technical title and when it likely would be applied or used in a game situation. After that introduction, the boy or girl demonstrates the entire skill in regular motion, much as one would demonstrate the entire operation in TWI's approach. Figure 8.16 shows a screenshot of this for the "Inside Turn," a move used when one wants to change directions quickly while dribbling the ball.

After that demonstration is completed, the boy or girl stops, and the narrator presents a step-by-step listing of what actions the skill or move involves. As the narrator identifies a given step or action, the step appears in text to the side of the image of the boy or girl. When the narrator finishes describing the step, the boy or girl performs that step and then stops to wait for the next step to be listed and described (Figure 8.17).

The same sequence is repeated until all steps have been presented and demonstrated; the entire operation is shown verbally (Figure 8.18). This is similar to the portion of TWI training wherein the trainer breaks down the task and talks through the important attributes of each step.

Figure 8.16. Initial demonstration: Full speed. All virtual soccer images provided by and used with permission of Challenger Sports Corporation.

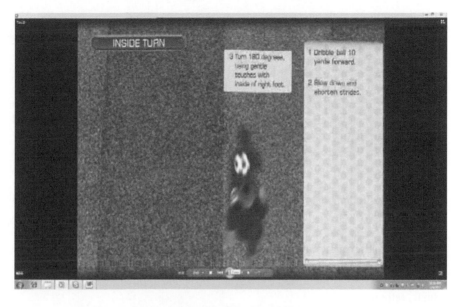

Figure 8.17. Breaking down step: Full Speed, but with stops.
All virtual soccer images provided by and used with permission of
Challenger Sports Corporation.

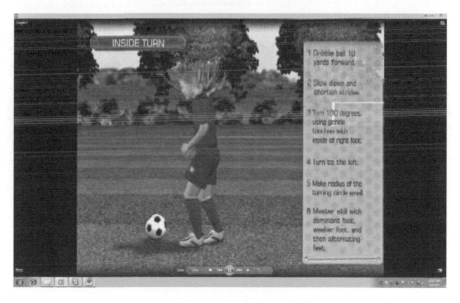

Figure 8.18. All steps shown in box.
All virtual soccer images provided by and used with permission of
Challenger Sports Corporation.

Next, the narrator introduces a slow motion version of the demonstration, and that is followed by the demonstration in regular motion. Both include the textbox showing the print-linguistic information associated with each step. Again, in TWI's program, one would slowly go through the steps and then let the trainee perform the steps. The slow motion demonstration allows the viewer to observe more closely the action with each aspect of the skill. The viewer observes the entire process and is able to see what it all looks like. This provides a visual representation that is comprehensive in the information it presents.

Throughout the segment, the visual and verbal information is separated so as not to force the viewer to process the different information at the same time. Further, the verbal information is provided two ways (textually and audio narration) to limit short-term memory dynamics to those two modes—visual (print-linguistic textual information) and audio. Providing it both ways also helps to reinforce that information. Ultimately, then, the viewer experiences the instructional information three ways: visual demonstration of the avatar, visually through the textual information, and aurally through the audio narration. This applies the principles identified by Mayer (2001) and applied by the TWI instructional method.

CONCLUSION

Much scholarship in technical communication, especially as related to multimodality, theorizes and encourages principles espoused in the TWI program in technical communication pedagogy. As their use in the workplace and pedagogy increases, students need to practice literacy skills related to producing multimodal products, ranging from literacy associated with using the technology itself to understanding its affordances and constraints to facilitate various operations and tasks in addition to understanding specific concepts in technical communication. However, these principles and skills are closely associated with those originally integrated in the TWI program. The sponsorship of TWI through the MEP and the pedagogies associated with multimodality encouraged through managerial and technical communication scholarship provide current examples of literacy sponsorship of its multimodal communication and related literacies.

CHAPTER 9

Workplace Communication and Implications of Sponsorship

> The capture of literacy for economic production and competition introduces great demand and support for writing yet also great instability and turbulence into workplace writing practices. People who write for a living must function under these conditions often as intense mediators of powerful ideological processes, mingling self and system as they transform abstract need into transactional texts.
>
> —Brandt, 2005, p. 194

Brandt's quotation above calls attention to the increasingly challenging relationship between ideologically motivated literacy learning and personal, economically motivated practices relative to valuations of literacy and related pedagogies. As workplaces strive to maximize profit in good economic times and survive in poor economic times, they tend to focus on short-term objectives. Individuals, however, look to fill short-term needs while balancing long-term needs that may affect their employment in dynamic economies. As institutions of sponsorship, workplaces have considerable power in dictating what literacy practices are valued and how they are valued. As such, the forms of writing practices that occur in workplaces tend to become part of literacy ideology. When the government supports certain technical communication literacy practices further, through explicit collaboration and coordination with industry, it reinforces the role that government plays as a sponsor as well. The reduction of literacy hierarchies within organizations toward literacy practices encountered in everyday activities facilitates better communication throughout the entire organization. This study has shown that when industry and government worked together to accommodate literacies with which workers were familiar, technical communication at the workplace improved. The bomb accident showed that a division in valuing practices could negatively impact communication efforts, so more materials that integrated both visual and print-linguistic attributes were developed at both levels—management and labor.

Graff (1979) asserted that any definition of literacy must be able to account for cultural and temporal differences in literacy dynamics, and Brandt (2001) called attention to changes in literacy standards over a period of time. While the field of literacy studies espouses research that considers ecological dynamics of literacy within social environments, it lacks close examinations of historical implications of certain forms of sponsorship. This study helps to address that gap.

As I acknowledge in Chapter 1, Street (1984) characterizes the meaning of literacy as embedded in the institutions in which it is practiced; a particular institution may value a certain form of literacy, and thus, that particular form of literacy is considered more appropriate than any other form. Brandt's (2001) quotation above echoes this attribute of literacy, as those institutions in which literacy is practiced sponsor those literacies that enable it to accomplish its work and economic goals at the risk of sacrificing long-term benefits of workers or neglecting individuals' needs. As Heath (1993) found in her study, children from the African American working-class community of Trackton were placed at a disadvantage in school compared with children from the White working-class community of Roadville because of differences in the ways literacy development occurred at home within each setting. African American children had little experience at home with the school-valued forms of literacy, while the White children had more experience with those literacies at home, and consequently, the African American children struggled in school.

This book has attempted to consider various literacy practices within a particular sponsorship dynamic that worked together at times and collided at other times with serious implications. This sponsorship dynamic starts in the workplace as a form of technical communication and extends into the homes, school, and community, suggesting the influence that the sponsorship of technical communication practices in the workplace has within the local ecology. While the primary research question associated with the historical study focused on the immediate workplace and community, broader questions associated with this study are How can workplaces and the government act more effectively to sponsor multiliteracies and multimodality as they are practiced in managerial and technical communication? Related questions include How can sponsorship of workplace literacy practices affect literacy practices within the community in which an employer operates, and how can an employer consider a community's literacy practices within its own literacy practices?

CONCLUSIONS ABOUT SPONSORSHIP AND MULTIMODAL LITERACIES

This study shows a clear link between literacy sponsorship and the impact sponsorship of technical communication has on a literacy ecology. Sponsorship of a particular literacy suggests an ideology that esteems that particular form of literacy. The government and operators of the Arsenal accommodated the literacies of the workers by providing training that emphasized visual, aural, and experiential practices and integrating visuals into instructional materials.

However, a clear emphasis on favoring print-linguistic literacy is evident in the hierarchy at the Arsenal; the adjusting of print-linguistic text to lower grade levels to accommodate low-literacy workers; and its presence in the school, at home, and in the community literacy practices. This emphasis suggests an ideology that esteems print-linguistic forms of literacy over other forms of literacy. Street (1984) asserts that literacy is defined and valued according to the institutions in which it is practiced, and the sponsors in this study valued print-linguistic literacies over others toward certain benefits.

Brandt and Clinton (2002) observed limitations that local studies of literacy like this one, which tend to focus on particular individuals or communities, have on understanding global issues. However, they also acknowledge that local practices tend to extend into more global settings. I identify and discuss several global implications associated with the findings of this study in the remainder of this chapter.

Literacy Sponsorship and Market Demands

Bazerman (2008) and Brandt (2001) note that literacy crises seem to be perceived at particular points in time. Brandt (2001) and Kress (2003) observe changes in what literacies are esteemed over time as different technologies are available and used. Brandt asserts that literacy crises emerge from changes in various geopolitical and global economic forces that shape the market demands for business and industry (2001, p 75). These in turn affect changes in literacy expectations to facilitate meeting market demands. Market demands include not only those literacies required to produce needed materials and products but also demands associated with sustaining a given social and economic structure and improving it.

During WWII, the government and plant operators needed to facilitate visual and experiential literacies because of the geopolitical and demographic factors associated with a grand scale, global war that took many skilled tradesmen out of industry and replaced them with people with different work-related skills. As evidenced in the interviews and examples of print materials reported here, lineworkers practiced predominantly visual and aural/oral literacy skills, while supervisors experienced a balance between the visual and the verbal forms of representation. Also, the higher-level administrators, who wrote the annual summaries of operations, emphasized print-linguistic skills. These differences in modes across the levels within the organization represent a form of social stratification. Workers who do not have certain literacy skills cannot advance beyond a given level, creating a literacy-oriented barrier to advancement. Literacy, then, is used to stratify the organization such that those who have esteemed print-linguistic skills have authority over those who lack those skills. If literacy practices across modes of representation are practiced such that all who experience them learn how to practice them with proficiency, that barrier is eliminated.

Workers are able to learn tasks more quickly when training integrates multi-modal forms of representation. Understanding factors that may affect worker training and the relationship between the visual and the textual modes of representation relative to a given context will help employers to develop appropriate training materials in such contexts while encouraging employees to advance their literacy skills. As such, this study contributes to an understanding of how workers' literacy backgrounds affect their ability to learn certain skills and how modes used in training practices can accommodate certain literacy backgrounds and experiences of new employees toward a more diversely literate workforce. It also encourages instruction in multimodal rhetoric and related literacies because of ways various modes can reinforce each other or serve to clarify information for readers.

Again, as the New London Group (1996) asserts, "The visual mode of representation may be much more powerful and closely related to language than 'mere literacy' would ever be able to allow" (p. 64). By emphasizing the visual skills of workers, the government and employers connected to the Arsenal effectively minimized literacy differences among workers, helping people from various backgrounds understand how to perform their specific tasks.

Because of the wartime economic environment, the operators of the Arsenal needed an efficient way to prepare workers for their tasks. Many migrants had skills training for farmwork, but they would need retraining for war-related industry. Training that emphasized the visual could accommodate faster transition in skills. This is further reinforced by the transient nature of the migrants. With a high turnover, employees needed to be able to be hired and placed to work quickly.

Also, interestingly, evidence presented in Chapter 6 suggests a correlation between level of employment and literacy skills required. Someone who had alphanumeric, print-text-based literacy skills was able to attain a higher level of employment because more documents at those levels emphasized print-linguistic modes of representation. Such skills were emphasized in the forms of representation used by those employees. Those who had visual and experiential literacy skills were limited to work at lower levels in the organization because training and practices emphasized those skills. However, the study also shows that various modes can act to reinforce each other. If these modes are not used in conjunction with each other, severe consequences can occur, as evidenced with the explosion accident.

Implications of Sponsorship Agency

Knowledge is delivered through literacy practices. Kucer and Silva (2006) note that literacy sponsors play a role in constructing knowledge for particular communities: "Literacy is a primary avenue through which knowledge is developed and conveyed. . . Sponsorship impacts what knowledge is to be privileged or deemed 'official' and what is to be ignored" (p. 43). Information needs of workers and others must consider the best literacy ways for the information to be presented. As the New London Group (1996) noted, certain modes of representation may serve readers better than other modes for particular information. This is also echoed in Moreno and Mayer's (2000) findings concerning use of multimedia forms of instruction to help students learn new concepts. The accident involving the bomb explosion illustrates the importance of having multiple forms of literacy practices reinforce each other to facilitate understanding and retention of information. Subsequent to the explosion, it is evident that employees received training about safety protocol as well as a handbook reinforcing that information.

The case discussed throughout Chapters 5–7 shows how the kind of literacy a sponsoring agent facilitates and encourages, while also regulating practices, can have drastic material consequences. Precautions that could have prevented the accidental explosion existed in a print-linguistic document available to higher-ranking officers of the Arsenal. While evidence that workers were aware that they needed to be careful with the boxes exists (the foreman acknowledging so), evidence also suggests they did not understand why or what precautions they should have taken (the foreman acknowledging that the boxes were still handled roughly). Had the information been available to them in print-linguistic form, or conveyed more precisely orally, the accident may have been avoided.

Changes brought on by war are sudden and require quick shifts in skills. Similarly, as technology changes at an increasingly faster pace and global competition intensifies, policy planners need to account for this potentially unanticipated shift and literacy issues therein. Sensitivity to such forces will help people understand why certain literacy practices are being valued at a given moment over others while recognizing potential value in other literacies. Further, in an economic downturn that includes increased unemployment rates in spite of reports of employers seeking qualified workers (Schleis, 2009), such as that experienced recently (Schleis, 2009), an environment that embraces multiple literacies may be able to adjust to changes in needed employee skills quickly. Such dynamics are the impetus for the MEP program's existence, further offering an illustration of sponsorship through technical communication practices.

By examining the literacy relationships that exist across the various settings of home, school, community, and work in a particular literacy ecology, a few observations may be inferred from this study. The first is that literacy-development policies must recognize that some geopolitical and geoeconomic forces that shape literacy demands cannot be planned for and require immediate shifts in literacy ideologies and practices to effectively negotiate changing political and economic patterns. The second is that multiple forms of literacy can be effectively practiced in a single ecology whether they are formally esteemed by a given entity or not, and these practices should not be ideologically prioritized. Third is that technical communication practiced in workplaces has a direct influence in sponsorship dynamics, suggesting that this role be part of technical communication education as an explicit link between technical communication and other literacy practices. By recognizing the value of multiple literacies and multimodal forms of representation, the government and workplaces can open doors to employment and opportunity when they encourage these practices to exist and value them.

Socioeconomic and Sociopolitical Implications

A few responses to these observations, some of which are already being espoused, include the following:

1. Multiple literacies, as encouraged in technical communication pedagogy, need to be valued, instead of valuing a single form of literacy.
2. Ecologies of literacy must account for sociopolitical, economic, and demographic dynamics at play in a given period of time and facilitate adaptation.
3. Sponsorship of literacy starts at the highest levels of government, business, and industry; and these entities need to value literacy development consistent with evolving global market needs.
4. These entities must support literacy development by making funding for multimodal and multiliteracy education more accessible.
5. Institutions in which literacy development occurs—schools, colleges, and workplaces—need to integrate work-related skill development into their curricula while facilitating disciplinary discourse learning; this may be facilitated via service learning and internship opportunities.

The use of multimodal representations and various representation systems individually can facilitate communication. However, the quote prefacing this chapter acknowledges that print-linguistic literacies are still valued in the workplace. Brandt's (2005) study pertained to 20th century skills and changes in print-linguistic literacies that occurred over that period. Whether these skills persist or change within the 21st century needs to be considered further. A growing body of scholarship observes a trend toward demands for visual literacies in the 21st century (Kress & Van Leeuwen, 2001; Mayer, 2001; Mitchell, 1995; Selfe, 2004; Tufte, 2006a, b). The federal government should fund education programs that support multimodal literacy development, much as it funded institutions that supported print-linguistic literacies in the period associated with this study and continues to fund multimodal technical communication practices through the MEP program.

Higher education seems to support disciplinary specialization as preached by business and industry leaders who influence federal education policy (again, the the Secretary of Education's Commission on the Future of Higher Education included a handful of business and industry executives). However, education needs to account for potential changes such that workers can shift from one kind of job to another as market demands change. This affects what institutions should teach to prepare future workers. Studies in workplace literacy support valuing multiple literacies equally (Kalyuga, 2005; Petroski, 1996; Senge, 2006; Witte, 1992; Zimmerman & Marsh, 1989).

While scholars like Olson (1993), Haas (1994), Jolliffe (1997) and Merrifield (1997) encourage discourse-specific training, Applebaum and Batt (1994), Gee, Hull, and Lankshear (1996), Dinero (2005), and Senge (2006) argue that the marketplace has changed significantly in the past 20 years such that competition is global and customers are more savvy (Applebaum & Batt, 1994; Gee et al., 1996). These attributes require workers to find creative ways to reduce costs, increase operating efficiencies, and to be able to customize products more for customers, requiring workers to be able to adapt quickly (Gee et al., 1996; Dinero, 2005; Senge, 2006).

Within this "new work order," Gee et al. (1996) identify metacognition as the primary discipline associated with the new worker; helping students understand how to do things within a given system and be creative problem-solvers. Dinero (2006) and Graupp and Wrona (2006) allude to creative thinking skills to help workplaces find more efficient methods for doing certain tasks. Further, Senge (2006) identifies five particular sets of skills, encouraging workers to constantly develop themselves so that they can adapt to new environments while also engaging in a systems approach to analysis: trying to understand how things operate within a given system and trying to improve certain attributes within it so that the rest of the system benefits. These skills and disciplines require various literacy practices identified above, including visual, experiential, and print-linguistic. While not a time of war, there is considerable geopolitical movement in business and industry that necessitates quick transitions. The current economic environment also demands that workers be able to transition from one set of skills to another quickly to minimize unemployment.

Implications for Workplaces and Literacy Education

Though this study examines practices and ecologies associated with a very unique period in this country's history—a period when the world was at war and most of the government, industry, and civilians' efforts were aimed at ending that war—this study's findings carry many implications pertaining to the question of how workplace literacy practices can affect literacy practices within the community in which an employer operates and how a community's literacy practices can affect an employer's literacy practices. Specifically, the following implications emerge:

1. Practices that encourage multiliteracies and multimodal communication facilitate rapid learning of new concepts, addressing language and literacy-related barriers.
2. There are connections between literacy practices and effective rhetoric. Certain literacies are more effective for certain purposes than are others, including instruction in new concepts; and these literacies ought to be taught and applied toward accomplishing those rhetorical purposes. This literacy includes an understanding of how to balance rhetoric associated with various modes of representation.
3. Economic support from the government and others can encourage development of proficiencies in multiliteracy practices that facilitate quick transitions from demands associated with one economy to those associated with another.

While certain rhetorical purposes affect literacy practices, as they consider how to train new employees, companies should conduct a needs assessment that addresses not just internal work dynamics but also the literacy backgrounds of their employees and employees' learning styles. Such assessment should also consider sociopolitical factors that affect the company's performance. Such assessment can help literacy sponsors such as the government, institutions of higher education, and companies understand how to facilitate efficient operations and transitions to new economic demands.

The current global economy encourages workplaces to relocate jobs to different parts of the world. As workers cross geographic and linguistic boundaries, their literacy skills are challenged. Multimodal forms of training facilitate learning in such environments. Workplaces also want employees to be able to transition quickly into their jobs without needing much training. As companies understand the training new-hires have received prior to being hired, they can use that background to help accommodate certain literacy skills and understand with which modes trainees feel they best learn. Further, facilitating multimodal forms of training will lessen the amount of time needed to train employees. It may also help workers who lack certain skills for other kinds of work learn those skills faster, helping them to move from one job to another faster.

Mitchell (1995) called attention to the increasingly global visual culture that includes flight simulators and computer-aided design tools and virtual environments in the workplace (p. 23). Also, recently, forms of training are being conducted via video gaming environments. Chapter 8 included several examples of these.

A point Gee (2003) makes about games and their ability to help people learn is that they need to be able to be learned. That is, learners need to be able to understand the game itself before they can use it for learning (p. 6). People need to have a certain kind of visual literacy in order to use games for learning (p. 13). Gee acknowledges that game environments represent a different "semiotic domain" than other, more commonly experienced semiotic domains. Gee espouses using video games for learning activities because of the social dimensions involved in such games, and their increasing popularity suggests that many people may be literate in their use.

Policymakers need to understand what skills are needed to accommodate shifts in demand as technology changes and demographics change, affecting economic change as well. Such skills are being identified in literature associated with the "new work order" (Gee et al., 1996) and "the learning organization" (Senge, 2006). Generally, these skills encourage proficiency in multiple literacies. The government is a major sponsor of literacy in both the local ecology associated with this study as well as in the national and global ecology.

The Training Within Industry (TWI) program that many arsenals used during WWII supported certain literacies and incorporated multiple modes of representation—aural, visual, and behavioral. Within educational settings, the TWI philosophy has been applied in service learning initiatives. Further, the philosophies developed by those who created the TWI program has been espoused recently in business and industry publications as one that can lead to more efficient production methods, and it has been implemented in several companies, including Toyota (Liker & Meier, 2007). This form of training is currently being supported as a form of efficient training in what Kress (2003) calls an increasingly visual culture. In such a culture, certain literacies are required. While it emphasizes visual and experiential literacies, TWI also encourages development of print-linguistic literacies. For example, such skills are integrated in its use of suggestion box practices and in creating narratives that not only may help in training but also in technological or process improvement (Dinero, 2005; Liker & Meier, 2007). As such, it integrates multiple literacies and engages various modes of representations associated with those literacies. However, with regard to this study, such print-linguistic literacies were not encouraged at lower levels of the organization, hurting employees.

The Training Within Industry model is still practiced, and an annual conference about the impact that it has had on companies has been held in recent years (TWI Summit). Further, recent publications such as *Toyota Talent* (Liker & Meier, 2007), *TWI Workbook* (Graupp & Wrona, 2006) and *Training Within Industry: The Foundation of Lean* (Dinero, 2005) espouse TWI as a training model. Such attention to modes used in training practices should include critical examination of political, cultural, and economic implications in addition to safety and risk concerns. The study I have reported here can help to improve connections between forms of workplace and community literacies and how companies can implement it and educational institutions can facilitate learning with it.

LIMITATIONS OF THIS STUDY

This study examines practices in a given ecology with the benefit of more historical perspective than one generally can ascertain in an ethnographic study. While this perspective offers a broader lens to facilitate analysis, several limitations of such historical studies are evident with this study. These limitations generally pertain to the inability to observe practices, having to rely on memories of events and practices that occurred 50 to 60 years ago, and the challenges of connecting available documentation with actual practices. I identified these issues in Chapter 2.

The most significant limitation is associated with the necessity to infer actual practices from interviews and documents, as opposed to observing those practices. As I acknowledged in Chapter 5, while there were several manuals/SOPs available for review in the archived materials, interviewees acknowledged very limited use of manuals in their training, if any use at all. Further, the report of the bomb explosion accident seems to indicate that even procedures presented in training were not followed strictly. They were modified to suit particular temporal concerns at the risk of safety. Actual observation of these practices would ascertain to what degree manuals were used along with any on-the-job training and how workers followed safety-related protocol.

Relative to inferences of home- and school-based practices, many interview participants rarely could recall specific practices, offering general recollections. For a study of general practices, like Brandt's (2001), such coverage is sufficient if accompanied by a few specific experiences to illustrate experiences those who could not recall specific events may have had. A few participants did identify specific activities, many of which I report in Chapters 5–7. This facilitated some inference to the general community's practices. Recollections of general practices at the Arsenal and the documents helped to infer practices there through triangulation. However, actual observation would enable the researcher to describe specific practices in considerably more detail than I could report here.

While observation would be optimal, a historical study that involved a more recent ecology, perhaps less than 15 years ago, could address some of these issues too. Interviewees' memories would be fresher and the participants would be younger, limiting lost recollections or foggy memories. I have attempted to address some of these limitations through triangulation of data sources and research methods, but I recognize that the limitations impact the ability to verify practices.

Another limitation, alluded to in Chapter 2, is that the sampling for the interview study included only Caucasians who grew up in the area or migrated and stayed in the area. It does not include African Americans who also migrated or grew up in the area and/or any people—African American or Caucasian—who relocated beyond the county. These challenges contribute to identifying attributes that future historical studies in technical communication can take.

DIRECTIONS FOR FUTURE RESEARCH

This study supports much of the current scholarship in literacy studies and technical communication scholarship pertaining to the advancement of the concept of multiple literacies and pedagogies that include instruction in various modes of representation, including multimodality. It calls attention to the separation of sponsored literacies and potential consequences of such separation. Future studies can consider these and other phenomena pertaining to literacy sponsorship within ecologies of literacy, including

1. Other historical studies on the subject of ecologies of literacy. For example, future studies can consider the impact that automotive manufacturing plants have had on certain locales, as the automotive industry changes directions within the current economy.
2. Historical studies, like this one, that also include the experiences of minority populations and those who migrated to other areas after their particular experience. Examination of government and military sponsorship regarding minority groups that served the country in wartime, and their experiences after war, can offer much for policy development that engages minority groups.
3. How the government supports and restricts literacy development when it sponsors a limited set of skills toward advancing what it perceives to be market trends. For example, the U.S. government offered federal grants supporting specific kinds of programs and literacies thereof. Studies can consider the impact this sponsorship has within historical contexts.
4. Considering how companies currently apply the Training Within Industry model and critiquing its successes or failures as a literacy sponsor.

I encourage further historical studies in literacy with the understanding that limitations such as these be recognized but that research into local practices with a historical perspective can benefit literacy research. As such studies are published, readers of these studies can attempt to ascertain ways to address these limitations.

ACTION PLAN

With the proliferation of new media, the concepts of multimodality and multimodal rhetoric have emerged as very important elements of literacy. Companies and individuals are communicating on a regular basis with others using digital media that allows for messages that are composed using multiple modes. Everyone needs to learn how to compose using these various modes and the tools that facilitate them. We also understand that certain modes tend to work better with each other than others relative to a given message and purpose. People need to learn how these modes work together toward particular purposes—the multimodal rhetoric of a message composed with these media.

I have tried to illustrate this rhetoric and relationships between modes in this text using a particular business model as a model of multimodal practices and show specific attributes of it with a detailed case study and several examples. However, a challenge that this text cannot address is the ever-evolving forms of new media that allow for various combinations of modes of representation.

As new media is invented and is refined, people need to learn additional attributes associated with those media that affect their rhetorical capabilities. Much as Tufte (2006a) did with PowerPoint, we need to identify limitations of tools that may seem powerful at first but show themselves not to be as effective as initially perceived. Part of literacy learning is understanding when not to use certain modes of representation or media tools associated with them because of limitations related to those modes and tools. This learning occurs by interacting regularly with the tools and messages, and that interaction occurs in education and work experiences, both of which integrate multimodal rhetorical elements of TWI.

The federal government, business entities, and schools are the most powerful sponsors of literacy development. As such, each bears some responsibility to facilitate multimodal literacies. Much as the federal government has funded the TWI program to encourage business to operate more efficiently, the government also needs to fund research into the study of multimodal rhetoric in addition to funding experiential learning initiatives. Businesses can also contribute to experiential learning programs by offering paid internships or internship-related scholarships. Educational institutions—secondary education and higher education—can develop cooperative agreements with businesses and governmental agencies to help fund education while also preparing workers to excel at a given job.

Whereas an internship or an entry-level training program used to fill the gap between graduation from high school or college and a steady job at a workplace, businesses are demanding graduates get that experience while in college or in a joint vocational-education setting. Typically, this means that college students are expected to have an internship experience at least by the time they are in their second semester of senior year, if not earlier. The student who is able to have multiple internship experiences in their junior and senior years is in a special position.

Most internships are volunteer experiences—they are unpaid jobs. While the company or entity benefits through the student's effort economically and materially, the student benefits only from the learning he or she experiences. Students are struggling to pay for college, and a paid internship may be an effective tool to balance the students' needs to learn through experience as well as the company's need to be profitable. Even if the internship pays minimum wage, it can be an effective learning tool as well as provide economic benefit to the company itself. Or perhaps companies can offer internship scholarships—paying for a student's tuition while that student works a semester for the company.

The government should also fund similar scholarships, putting students to work in an internship capacity for a government agency at the local, state, or federal level. Such support benefits government agencies by putting well-educated people to work in those agencies while also benefiting students by helping to pay for their education.

Educational institutions are requiring students to get experience with certain job skills in order to graduate. These may be part of a service initiative or internship program. In either case, the institution needs to establish relationships with as many entities as it can in its local area or beyond if telecommuting work is possible.

The government, businesses, and educational institutions bear responsibility for ensuring this multimodal rhetorical learning occurs and facilitates economic development.

CONCLUSION

My hope with this book was to raise awareness of the role technical communication plays as a sponsor of literacy practices. As new media is developed and new modal combinations to communicate emerge, technical communication is at the cutting edge of literacy. Technical communication is sponsored by industry and government in a variety of ways, and that sponsorship impacts literacy education at several levels. For example, technical communication coursework is explicitly integrated into engineering programs.

TWI serves as an example of the link between the technical communication sponsorship dynamics of WWII and today. The multimodal rhetoric espoused in TWI is part of technical communication scholarship as well as educational psychology scholarship; people are able to communicate effectively using multiple modes of representation, and the use of multiple modes of representation in instruction facilitates better learning. Different modal combinations affect cognition, and scholarship is trying to ascertain the range of possible combinations. Technical communication scholarship has come from recognizing graphics as "additive" (Debs, 1988) to developing an entire subfield in visual and multimodal rhetoric.

However, government also influences literacy learning by providing scholarships and grants to support certain practices and programs. Consequently, government and industry have a responsibility to facilitate multimodal communication learning, much as they joined forces to develop TWI and MEP to improve operating efficiencies. Much as TWI experienced change toward improving its rhetoric, so too must technical communication scholarship report on developments in multimodal designs that improve communication.

By examining historical practices in their context and linking those practices with current practices and scholarship, researchers and teachers can facilitate better sponsorship of technical communication from industry and government through education and in education.

References

Amerine, R., & Bilmes, J. (1990). Following instructions. In M. Lynch, & S. Woolgar (Eds.), *Representation in scientific practice* (pp. 323–336). Cambridge, MA: MIT Press.

Ammunition, General, TM 9-1900 War Department Technical Manual. (1945). Washington, DC: Government Printing Office.

Applebaum, E., & Batt, R. (1994). *The new American workplace: Transforming work systems in the United States*. Ithaca, NY: ILR.

Army Service Forces, Office of the Chief of Ordnance, Office of the Field Director of Ammunition Plants. (1945). *Standard practice manual 250 LB. Bomb, G.P., AN-M57A1 500 LB. Bomb, G.P., AN-M64A11000 LB. Bomb, G.P., AN-M65A1 2000 LB. Bomb, G.P., AN-M66A1 Tritonal Loaded.*

Arnheim, R. (1969). *Visual thinking*. Berkeley, CA: University of California Press.

Atlas Powder Company. (1942). *Semiannual summary of operations: Boomtown Army ammunition plant (AAP)*. AAP Repository Documents.

Atlas Powder Company. (1944a). *History of the operating contractor's organization and operation of the Boomtown ordnance plant* (Vol. 1). Wilmington, DE: Government Printing Office.

Atlas Powder Company. (1944b). *History of the operating contractor's organization and operation of the Boomtown ordnance plant* (Vol. II). Wilmington, DE: Government Printing Office.

Atlas Powder Company. (1944c). *Semiannual summary of operations: Army ammunition plant (AAP)*. AAP Repository Documents.

Atlas Powder Company. (1945). *Semiannual summary of operations: Ravenna Army ammunition plant (AAP)*. AAP Repository Documents.

Baddeley, A. D. (1986). *Working memory*. Oxford, NY: Oxford University Press.

Barr-Capper, L. G. (2006). *Appalachian women's stories of migration from West Virginia to northeast Ohio: A narrative analysis of interviews and personal observations*. Retrieved October 11, 2006, from http://www.marshall.edu/csega/research/linda capper.pdf

Barton, D. (2007). *Literacy: An introduction to the ecology of written language*. Malden, MA: Blackwell.

Bateman, J., Delin, J., & Henschel, R. (2007). Mapping the multimodal genres of traditional and electronic newspapers. In T. D. Royce & W. L. Bowcher (Eds.), *New directions in the analysis of multimodal discourse* (pp. 147–172). Mahwah, NJ: Lawrence Erlbaum.

Bazerman, C. (2008). Theories of the middle range in historical studies of writing practice. *Written Communication, 25,* 298–318.

Benson Polytechnic High School. (2006). *Expansion and the second world war.* Retrieved October 10, 2006, from http://en.wikipedia.org/wiki/benson_polytechnic_high_school

Brandt, D. (1999). *Literacy, opportunity, and economic change.* Albany, NY: National Research Center on English Learning & Achievement.

Brandt, D. (2001). *Literacy in American lives.* Cambridge, MA: Cambridge University Press.

Brandt, D. (2005). Writing for a living: Knowledge and the knowledge economy. *Written Communication, 22,* 166–197.

Brandt, D., & Clinton, K. (2002). Limits of the local: Expanding perspectives on literacy as a social practice. *Journal of Literacy Research, 34,* 337–356.

Brumberger, E. (2008). Visual communication in the workplace. *Technical Communication Quarterly, 16,* 369–399.

Cicourel, A. V. (1981). Language and medicine. In C. A. Ferguson & S. B. Heath (Eds.), *Language in the USA* (pp. 407–429). Cambridge, MA: Cambridge University Press.

Cooper, M. M. (1986). The ecology of writing. *College English, 48,* 364–375.

Crano, W. D., & Brewer, M. B. (2002). *Principles and methods of social research* (2nd ed.). Mahwah, NJ: Lawrence Erlbaum.

Debs, M. B. (1988). A history of advice: What experts have to tell us. In E. Doheny-Farina (Ed.), *Effective documentation: What we have learned from research* (pp. 11–24). Cambridge, MA: MIT Press.

Decker, C. L., & Adamek, M. E. (2004). Meeting the challenges of social work research in long term care. *Social Work in Health Care, 38,* 47–65.

Dewey, J. (1938). *Experience and education.* New York, NY: Collier.

deWinter, J., & Vie, S. (2008). Press enter to "say": Using *Second Life* to teach critical media literacy. *Computers and Composition, 25,* 313–322.

Dinero, D. A. (2005). *Training Within Industry: The foundation of lean.* New York, NY: Productivity.

Dooley, C. R. (1945a). *Training within industry report 1940–1945.* Washington, DC: Government Printing Office.

Dooley, C. R. (1945b). "Preface." *Training within industry report 1940–1945.* Washington, DC: Government Printing Office.

Eeles, E., & Rockwood, K. (2008). Delirium in the long term care setting: Clinical and research challenges. *Journal of American Medical Directors Association, 9,* 157–161.

Establishment of the Boomtown Arsenal. (1943). Washington DC: Government Printing Office.

Feathers, C. E. (1998). *Mountain people in a flat land: A popular history of Appalachian migration to northeast Ohio, 1940–1965.* Athens, OH: Ohio University Press.

Fleckenstein, K. S., Spinuzzi, C., Rickly, R. J., & Papper, C. C. (2008). The importance of harmony: An ecological metaphor for writing research. *College Composition and Communication, 60,* 388–419.

Gaillet L. L. (2012). (Per)Forming archival research methodologies. *College Composition and Communication, 64,* 35–58.

Gee, J. P. (1996). *Social linguistics and literacies: Ideology in discourses* (2nd ed.). London, UK: RoutledgeFalmer.

Gee, J. P. (2003). *What video games have to teach us about learning and literacy.* New York, NY: Palgrave Macmillan.

Gee, J. P., Hull, G., & Lankshear, C. (1996). *The new work order: Behind the language of the new capitalism.* Boulder, CO: Westview.

Geisler, C. (2004). *Analyzing streams of language: Twelve steps to the systematic coding of texts, talk, and other verbal data.* New York, NY: Pearson.

Grabill, J. (2001). *Community literacy and the politics of change.* Albany, NY: SUNY Press.

Graff, H. (1979). *The literacy myth: Literacy and social structure in the nineteenth-century city.* New York, NY: Academic.

Graff, H. J. (2003). Introduction to historical studies of literacy. *Interchange, 34,* 123–131.

Graupp, P., & Purrier, M. (2012). *Getting to standard work in health care: Using TWI to create a foundation of quality care.* New York, NY: Productivity.

Graupp, P., & Wrona, R. J. (2006) *The TWI workbook: Essential skills for supervisors.* New York, NY: Productivity.

Gurak, L. J., & Lannon, J. M. (2007). *A concise guide to technical communication* (3rd ed.). New York, NY: Bedford.

Haas, C. (1994). Learning to read biology: One student's rhetorical development in college. In E. Cushman, E. R. Kintgen, B. M. Kroll, & M. Rose (Eds.), *Literacy: A critical sourcebook* (pp. 358–375). Boston, MA: Bedford/St. Martin's.

Heath, S. B. (1993). The madness(es) of reading and writing ethnography. *Anthropology and Education Quarterly, 24,* 256–268.

Heath, S. B. (2007). *Ways with words: Language, life and work in communities and classrooms.* Cambridge, MA: Cambridge University Press. (Original work published 1983)

Helmers, M. (2006). *The elements of visual analysis.* New York, NY: Pearson.

Hobbs, S. D. (1998). Foreword. In C. E. Feathers (Ed.), *Mountain people in a flat land: A popular history of Appalachian migration to northeast Ohio, 1940–1965.* Athens, OH: Ohio University Press.

Holsti, O. R. (1969). *Content analysis for the social sciences and humanities.* Reading, PA: Addison-Wesley.

Hudson, K. (2010). Applied training in virtual environments. In W. Ritke-Jones (Ed.), *Virtual environments for corporate education: Employee learning and solutions* (pp. 110–122). Hershey, PA: IGI Global.

Hull, G. (1997). Hearing other voices: A critical assessment of popular views on literacy and work. In E. Cushman, E. R. Kintgen, B. M. Kroll, & M. Rose (Eds.), *Literacy: A critical sourcebook* (pp. 660–684). Boston, MA: Bedford/St. Martin's.

Hull, G., & Schultz, K. (2002). Introduction: Negotiating the boundaries between school and non-school literacies. In S. Hull & K. Schultz (Eds.), *School's out: Bridging out-of-school literacies with classroom practice* (pp. 1–10). New York, NY: Teachers College Press.

Huot, B., Strobel, B., & Bazerman, C. (Eds.). (2004). *Multiple literacies for the 21st century.* Cresskill, NY: Hampton.

Huot, B., & Strobel, B. (2004). Introduction. In B. Huot, B. Strobel, & C. Bazerman (Eds.), *Multiple literacies for the 21st century* (pp. 1–12). Cresskill, NY: Hampton.

Hutchins, E. (1995). *Cognition in the wild.* Cambridge, MA: MIT Press.

Index, Boomtown Ordnance Center, Reduced Line Layouts. (1945). 12-13-44, Rev. 4-25-45.

Johnson, C. S. (2009). *The language of work: Technical communication at Lukens Steel, 1810 to 1925.* Amityville, NY: Baywood.

Johnson, S. A. (2006). *Industrial voyagers: A case study of Appalachian migration to Akron, Ohio: 1900–1940.* (Electronic Thesis or Dissertation.) Accessed February 20, 2008 from https://etd.ohiolink.edu/

Johnson, T. (2008). How to create video tutorials. *I'dratherbewriting.com.* Retrieved January 10, 2009, from http://www.idratherbewriting.com/2008/09/11/how-i-create-video-tutorials/

Johnson-Sheehan, R. (2012). *Technical communication today* (4th ed.). Boston, MA: Pearson.

Jolliffe, D. (1997). Finding yourself in the text: Identity formation in the discourse of workplace documents. In G. Hull (Ed.), *Changing work, changing workers: Critical perspectives on language, literacy, and skills* (pp. 335–349). Albany, NY: State University of New York Press.

Jones, J. (2007, April 23). The necessity of teaching video composition. *Viz. Visual rhetoric – visual culture – pedagogy.* Retrieved November 20, 2009, from http://viz.cwrl.utexas.edu/taxonomy/term/34

Kalyuga, S. (2005). Prior knowledge principle in multimedia learning. In R. E. Mayer (Ed.), *The Cambridge handbook of multimedia learning* (pp. 325–336). Cambridge, MA: Cambridge University Press.

Kirkevold, M., & Bergland, A. (2007). The quality of qualitative data: Issues to consider when interviewing participants who have difficulties providing detailed accounts of their experiences. *International Journal of Qualitative Studies on Health and Well-Being, 2,* 68–75.

Kolin, P. C. (2009). *Successful writing at work* (2nd ed.). Boston, MA: Houghton Mifflin.

Kress, G. (2003). *Literacy in the new media age.* London, UK: Routledge.

Kress, G., & van Leeuwen, T. (1996; 2006). *Reading images: The grammar of visual design.* London, UK: Routledge.

Kress, G., & van Leeuwen, T. (2001). *Multimodal discourse: The modes and media of contemporary communication.* London, UK: Arnold.

Krippendorff, K. (2004). *Content analysis: An introduction to its methodology* (2nd ed.). Thousand Oaks, CA: Sage.

Krippendorff, K. (2008). Reliability issues in content analysis data: What it is and why. In K. Krippendorff & A. Bock (Eds.), *The content analysis reader* (pp. 350–357). Thousand Oaks, CA: Sage.

Kucer, S. B., & Silva, C. (2006). *Teaching the dimensions of literacy.* Mahwah, NJ: Lawrence Erlbaum.

Liker, J., & Meier, D. (2007). *Toyota talent: Developing your people the Toyota way.* New York, NY: McGraw-Hill.

Linden Labs. (2010). Machinima content companies. *Second Life.* Retrieved January 21, 2011, from http://wiki.secondlife.com/wiki/Machinima_Content_Companies

Lynott, L. (1989, May 9). Apple Grove face-lift key to revival. *Record Courier* (Kent, Ohio).

Mackinnon, J., & Towell, C. (2009, May 15). End of the road for local dealers. *Akron Beacon Journal,* p A1, A10.

MacNealy, M. S. (1999). *Strategies for empirical research in writing.* Boston, MA: Allyn and Bacon.

Manual of Safe Practices. (1943). Washington DC: Government Printing Office.

Markel, M. (2010). *Technical communication* (9th ed.). Boston, MA: Bedford/St. Martin's.

Matthiessen, C. M. I. M. (2007). The multimodal page: A system functional exploration. In T. D. Royce & W. L. Bowcher (Eds.), *New directions in the analysis of multimodal discourse* (pp. 1–62). Mahwah, NJ: Lawrence Erlbaum.

Mayer, R. E. (2001). *Multi-media learning.* Cambridge, MA: Cambridge University Press.

Mayer, R. E. (Ed.). (2005). *The Cambridge handbook of multimedia learning.* Cambridge, MA: Cambridge University Press.

Merrifield, J. (1997). If job training is the answer, what's the question? In G. Hull, (Ed.), *Changing work, changing workers: Critical perspectives on language, literacy, and skills* (pp. 273–294). Albany, NY: SUNY Press.

Mesko, B. (2008, August 17). Unique medical simulation in Second Life. *ScienceRoll.* [Weblog post]. Retrieved May 20, 2012, from http://scienceroll.com/2008/08/17/unique-medical-simulation-in-second-life/

Miller, D. (Ed.) (2005). Going lean in healthcare. Cambridge: Institute for Healthcare Improvement. Accessed April 20, 2012 from http://www.entnet.org/Practice/upload/Going LeaninHealthCareWhitePaper.pdf

Mishra, S., & Sharma, R. C. (2005). *Interactive multimedia in education and training.* Hershey, PA: Idea Group.

Mitchell, W. J. T. (1995). *Picture theory.* Chicago, IL: University of Chicago Press.

Mollman, S. (2007, July 27). Wii + Second Life = new training simulator. *Wired.* Retrieved December 31, 2011, from http://www.wired.com/gadgets/miscellaneous/news/2007/07/wiimote

Moreno, R., & Mayer, R. E. (2000). A learner-centered approach to multimedia explanations: Deriving instructional design principles from cognitive theory. *Interactive Multimedia Electronic Journal of Computer-Enhanced Learning, 2.* Accessed April 20, 2008 from http://imej.wfu.edu/Aricles/2000/2/05/index.asp

Murray, J. (2009). *Non-discursive rhetoric: Image and affect in multimodal composition.* New York, NY: SUNY Press.

Nelson, D. (1995). *Farm and factory: Workers in the midwest 1880–1990.* Bloomington, IN: Indiana University Press.

New London Group. (1996). A pedagogy of multiliteracies: Designing social futures. *Harvard Educational Review, 66,* 60–92.

Northey, M. (1990). The need for writing skill in accounting firms. *Managerial Communication Quarterly, 3,* 474–495.

Ohio History Central. (2008). *Boomtown arsenal.* Retrieved June 25, 2008, from http://www.ohiohistorycentral.org

Oliu, W. E., Brusaw, C. T., & Alred, G. J. (2010). *Writing that works* (10th ed.). Boston, MA: Bedford/St. Martin's.

Olsen, L. A. (1993). Research on discourse communities: An overview. In R. Spilka (Ed.), *Writing in the workplace: New research perspectives* (pp.181–194). Carbondale,: IL Southern Illinois University Press.

Ordnance School Manual. (1941). Washington, DC: U.S. Government Printing Office.

Petroski, H. (1996). *Invention by design: How engineers move from thought to thing.* Cambridge, MA: Harvard University Press.

Pfingsten, R. (2010). *The history of the Boomtown Arsenal.* Lakewood, OH: Adkins.

Phelps, L. W. (1991). Practical wisdom and the geography of knowledge in composition *College English, 53,* 863–885.

Pinker, S. (1997). *How the mind works.* New York, NY: Norton.

Porter, J. E., Sullivan, P., Blythe, S., Grabill, J. T., & Miles, L. (2000). Institutional critique: A rhetorical methodology for change. *College Composition and Communication, 51,* 610–642.

Prufer, T. (1982). *History of the Windham schools—A sociological perspective.* Academic Paper.

Purcell-Gates, V. (1995). *Other people's words: The cycle of low literacy.* Cambridge, MA: Harvard University Press.

Remley, D. (2009). Training Within Industry as short-sighted community literacy-appropriate training program: A case study of a worker-centered training program. *Community Literacy Journal, 3*(2), 93–114.

Remley, D. (2010a). Developing digital literacies in *Second Life*: Bringing *Second Life* to business writing pedagogy and corporate training. In W. Ritke-Jones (Ed.), *Handbook of research on virtual environments for corporate education: Employee learning and solutions* (pp. 169–193). Hershey, PA: IGI Global.

Remley, D. (2010b). Second Life literacies: Critiquing writing technologies of Second Life. *Computers and Composition Online.* Retrieved from http://www.bgsu.edu/cconline/Remley/

Remley, D. (2012). Forming assessment of machinima video. *Computers and Composition Online.* Retrieved from http://www.bgsu.edu/cconline/cconline_Sp_2012/SLassesswebtext/index.html

Rodabaugh, J. H. (1975). Ohio and World War II. In T. Smith (Ed.), *An Ohio reader: Reconstruction to the present* (pp. 314–318). Grand Rapids, MI: William B. Eerdman.

Rowley-Joliet, E. (2004). Different visions, different visuals: A social semiotic analysis of field—Specific visual composition in scientific conference presentations. *Visual Communications, 3*(2), 145–177.

Royce, T. D. (2007). Intersemiotic complementarity: A framework for multimodal discourse analysis. In T. D. Royce & W. L. Bowcher (Eds.), *New directions in the analysis of multimodal discourse* (pp. 63–109). Mahwah, NJ: Lawrence Erlbaum.

Schacht, W. H. (2009). Manufacturing extension partnership program: An overview. *Congressional Research Service.* Retrieved January 26, 2012, from http://www.dtic.mil/cgi-bin/GetTRDoc?AD=ADA502057

Schleis, P. (2009, May 9). Unemployed lack skills for new jobs. *Akron Beacon Journal,* p. A6.

Schnotz, W. (2005). An integrated model of text and picture comprehension. In R. E. Mayer (Ed.), *The Cambridge handbook of multimedia learning* (pp. 49–60). Cambridge, MA: Cambridge University Press.

Scribner, S., & Cole, M. (1981). *The psychology of literacy.* Cambridge, MA: Harvard University Press.

Secretary of Education's Commission on the Future of Higher Education. (2006). *A test of leadership: Charting the future of U.S. higher education.* Washington, DC: Government Printing Office.

Selfe, C. (2004). Toward new media text: Taking on the challenges of visual literacy. In A. F. Wysocki, J. Johnson-Eilola, C. Selfe, & G. Sirc (Eds.), *Writing new media: Theory and applications for expanding the teaching of composition* (pp. 67–110). Logan, UT: Utah State University Press.

Senge, P. M. (2006). *The fifth discipline.* New York, NY: Doubleday.

Shrestha, L. B. (2006). *Life expectancy in the United States.* CRS Report for Congress. Washington, DC: Congressional Research Service.

Silverman, D. (2006). *Interpreting qualitative data* (3rd ed.). London, UK: Sage.

Staggers, J. M. (2006). Learning to love the bomb: Secrecy and denial in the atomic city, 1943–1961. Doctoral dissertation, Purdue University, 2006. *Dissertation Abstracts International, 67*(09). (UMI No. AAT 3232241)

Stratton, R. C. (1943, August 13). *Explosion while handling fuzed twenty-pound fragmentation bomb clusters* [Correspondence report].

Strauss, A. L. (1987). *Qualitative analysis for social scientists.* Cambridge, MA: Cambridge University Press.

Street, B. (1984) *Literacy in theory and practice.* Cambridge, MA: Cambridge University Press.

Supervisor Manual. (1945). Washington, DC: Government Printing Office.

Szwed, J. F. (1981). The ethnography of literacy. In E. Cushman, E. R. Kintgen, B. M. Kroll, & M. Rose (Eds.), *Literacy: A critical sourcebook* (pp. 421–429). Boston, MA: Bedford/St. Martin's.

Taub, H. A. (1980). Informed consent memory and age. *The Gerontologist, 20,* 686 690.

Taylor, F. W. (1913). *The principles of scientific management.* New York, NY: Harper & Brothers [E-book]. Retrieved September 28, 2012, from http://books.google.com/books?id=HoJMAAAAYAAJ&pg=PA3#v=onepage&q&f=false

Tri-county regional planning commission (1963). *Socio-economic Study: Boomtown, Ohio.* Akron: Community Assistance Division.

Troyer, L. (1998). *Portage pathways.* Kent, OH: Kent State University Press.

Tufte, E. R. (2006a). *The cognitive style of PowerPoint: Pitching out corruption within.* Cheshire, CT: Graphics.

Tufte, E. R. (2006b). *Beautiful evidence.* Cheshire, CT: Graphics.

TWI Institute. (2012). *Training within industry in healthcare.* Retrieved January 12, 2012, from http://twi-institute.com/healthcare_main.htm

TWI Learning Partnership. (2009). Retrieved from http://www.twilearningpartnership.com

Unsworth, L. (2007). Multiliteracies and multimodal text analysis in classroom work with children's literature. In T. D. Royce & W. L. Bowcher (Eds.), *New directions in the analysis of multimodal discourse* (pp. 331–359). Mahwah, NJ: Lawrence Erlbaum.

UW-Extension. (2006). *History of UW-Extension: 1940–1960: World War II and the post war boom.* Retrieved October 10, 2006, from http://www.uwex.edu/about/history

van Leeuwen, T. (2003). A multimodal perspective on composition. In T. Ensink & C. Sauer (Eds.), *Framing and perspectivising in discourse.* Amsterdam, The Netherlands: Benjamins.

Vie, S. (2008a). Tech writing, meet *Tomb Raider*: Video and computer games in the technical communication classroom. *E-Learning and Digital Media.* Retrieved January 19, 2011, from http://www.wwwords.co.uk/rss/abstract.asp?j=elea&aid=3318

Vie, S. (2008b). Are we really worlds apart? Building bridges between Second Life and Secondary Education. *Computers and Composition Online.* Retrieved December 10, 2008, from http://www.bgsu.edu/cconline/gaming_issue_2008/Vie_Second_Life/index.html

Wagner, M. (2007). Using Second Life as a business-to-business tool. *InformationWeek; The Business Value of Technology.* Retrieved September 10, 2010, from http://www.informationweek.com/blog/main/archives/2007/04/using_second_li_2.html;jsessionid=KIHDIBAKMQZBDQE1GHPSKH4ATMY32JVN

Walsh, R. (1995). *The World War II Ordnance Department's Government-Owned Contractor-Operator (GOCO) industrial facilities: Boomtown ordnance plant historic investigation.* Fort Worth, TX: Army Corps of Engineers.

War Manpower Commission. (1945). *Training Within Industry report 1940–1945.* Washington, DC: Government Printing Office.

Wattenburg, B. (2006). *FMC program segments 1930–1960: World War II.* Retrieved October 16, 2006, from http://www.pbs.org/fmc/segments/progseg8.htm

Weber, R. P. (1990). *Basic content analysis* (2nd ed.). Newbury Park, CA: Sage.

Winsor, D. (2000). Ordering work: Blue-collar literacy and the political nature of genre. *Written Communication, 17,* 155–184.

Witte, S. P. (1992). Context, text, intertext: Toward constructivist semiotic of writing. *Written Communication, 9,* 237–308.

Wysocki, A. F. (2001). Impossibly distinct: On form/content and word/image in two piece computer-based interactive media. *Computers and Composition, 18,* 207–234.

Yin, R. K. (1984). *Case study research: Design and methods.* Newbury Park, CA: Sage.

Zimmerman, M., & Marsh, H. (1989). Storyboarding and industrial proposal: A case study of teaching and producing writing. In C. B. Matalene (Ed.), *Worlds of writing: Teaching and learning discourse communities of the work* (pp. 203–221). New York, NY: Random House.

Index